I0488632

NUREG–1407

Procedural and Submittal Guidance for the Individual Plant Examination of External Events (IPEEE) for Severe Accident Vulnerabilities

Final Report

Manuscript Completed: April 1991
Date Published: June 1991

J. T. Chen, N. C. Chokshi, R. M. Kenneally, G. B. Kelly,
W. D. Beckner, C. McCracken, A. J. Murphy, L. Reiter, D. Jeng

Division of Safety Issue Resolution
Office of Nuclear Regulatory Research
U.S. Nuclear Regulatory Commission
Washington, DC 20555

ABSTRACT

Based on a Policy Statement on Severe Accidents, the licensee of each nuclear power plant is requested to perform an individual plant examination. The plant examination systematically looks for vulnerabilities to severe accidents and cost-effective safety improvements that reduce or eliminate the important vulnerabilities. This document presents guidance for performing and reporting the results of the individual plant examination of external events (IPEEE). The guidance for reporting the results of the individual plant examination of internal events (IPE) is presented in NUREG-1335.

CONTENTS

Appendices

Figure

Tables

EXECUTIVE SUMMARY

Background

In the Commission policy statement on severe accidents in nuclear power plants issued on August 8, 1985, the Commission concluded, based on available information, that existing plants pose no undue risk to the public health and safety and that there is no present basis for immediate action on any regulatory requirements for these plants. However, the Commission recognized, based on NRC and industry experience with plant-specific probabilistic risk assessments (PRAs), that systematic examinations are beneficial in identifying plant-specific vulnerabilities to severe accidents that could be fixed with low-cost improvements. As part of the implementation of the Severe Accident Policy, the Commission issued Generic Letter 88-20 on November 23, 1988, requesting that each licensee conduct an individual plant examination (IPE) for internally initiated events including internal flooding.

Many PRAs indicate that, in some instances, the risk from external events could contribute significantly to core damage. However, the examination for externally initiated events is proceeding on a later schedule to allow the staff to carry out additional work to (1) identify which external hazards need a systematic examination, (2) identify acceptable examination methods and develop procedural and submittal guidance, and (3) coordinate the individual plant examination of external events (IPEEE) with other ongoing external event programs. In December 1987, an External Events Steering Group (EESG) was established to make recommendations regarding the scope, methods, and coordination of the IPEEE. The EESG completed its task in May 1990. Based on the EESG recommendations, the staff prepared this report to provide detailed guidance to the licensees on the conduct of the IPEEE and on the structure and content of the IPEEE submittal. The staff issued a draft of this report for public comment in July 1990. It held a workshop in September 1990 to explain the IPEEE process and to obtain specific comments and questions on draft Supplement 4 to Generic Letter 88-20 and the draft of this document). In addition to numerous comments and questions raised during the workshop, the staff received written comments from 16 organizations.

This revised report reflects the staff's thorough consideration of the public comments received. It provides, specifically, the guidelines defining the IPEEE objectives; identifies external events that should be included in the IPEEE; identifies acceptable methodologies; identifies coordination between the IPEEE and the ongoing NRC programs; and provides the staff's responses to public comments and questions.

Objectives of the IPEEE

The general objectives of the IPEEE are similar to that of the IPE—that is, for each licensee (1) to develop an appreciation of severe accident behavior, (2) to understand the most likely severe accident sequences that could occur at its plant under full-power operating conditions, (3) to gain a qualitative understanding of the overall likelihood of core damage and fission product releases, and (4) if necessary, to reduce the overall likelihood of core damage and fission product releases by modifying, where appropriate, hardware and procedures that would help prevent or mitigate severe accidents. The key outcome of an IPEEE is the knowledge and appropriate improvements resulting from such an examination process. The examination can be conducted using any of the acceptable approaches.

Identification of External Events Included in the IPEEE

In supporting the implementation of the Severe Accident Policy, a study was performed to determine which external initiators could be a potentially important accident initiator that may pose a threat of severe core damage or of a large radioactive release to the environment. The external events considered, consistent with past probabilistic risk assessments (PRAs), are those events whose cause is external to all systems used during normal and emergency operations. The external events evaluated include seismic events, internal fires, high winds and tornadoes, external floods, transportation and nearby facility accidents, lightning, severe temperature transients (extreme heat, extreme cold), severe winter storms, external fires (forest fires, grass fires), extraterrestrial activity (meteorite strikes, satellite falls), and volcanic activity.

Based on the results of that study, the staff has concluded that five external events need to be included specifically in the IPEEE: seismic events, internal fires, high winds, floods, and transportation and nearby facility accidents. However, licensees should confirm that no other plant-unique external events with potential severe accident vulnerability are being excluded from the IPEEE.

Examination Methods

Seismic Events

A seismic PRA (Level 1 plus containment performance analysis) or a seismic margins methodology (SMM) is considered a viable approach to identify potential vulnerabilities. Guidance is provided for licensees performing a new seismic PRA or updating an existing seismic PRA;

emphasis is placed on the identification and ranking of dominant plant sequences that could lead to seismically induced core damage rather than on the numerical estimate of absolute frequency of occurrence. Methodology upgrades include plant walkdowns, evaluation of relay chatter, and evaluation of the effects of soil liquefaction.

Guidance is also provided for licensees using either the NRC- or EPRI-sponsored seismic margins methodology. The margins methodology screens components according to their importance to safety and seismic capacity. By design, the methodology utilizes two review or screening levels geared to peak ground accelerations of 0.3g and 0.5g. Review level earthquakes (RLEs) were assigned based on the Lawrence Livermore National Laboratory (LLNL) and Electric Power Research Institute (EPRI) hazard estimates, sensitivity tests, seismological and engineering judgment, and plant design considerations. The use of the 0.3g full-scope and focused-scope RLE for most plants in the Central and Eastern United States would meet IPEEE objectives. The level of effort in the analysis of relay chatter is the major difference between these two categories. For some sites where the seismic hazard is low, a reduced-scope margins methodology emphasizing plant walkdowns is considered adequate. For sites in the Western United States, except California coastal sites, the 0.5g RLE should be used. Methodology upgrades include relay chatter, liquefaction, and plant walkdown enhancements for the NRC method; guidance on alternative success paths for the EPRI method; and evaluation of nonseismic failure and human actions for both methods.

Internal Fires

The internal fires IPEEE can be accomplished by performing a Level 1 fire PRA. Those issues identified in the Fire Risk Scoping Study (NUREG/CR-5088) should be addressed using plant-specific data and a specially tailored walkdown procedure.

The guidance does not address the fire vulnerability evaluation (FIVE) methodology currently being developed by the Nuclear Management and Resources Council (NUMARC) and EPRI. This methodology is being reviewed by the staff; when the review is completed, the staff will issue an evaluation report on its acceptability for use in the IPEEE.

High Winds, Floods, and Transportation and Nearby Facility Accidents

The recommended overall approach consists of a progressive screening. The screening criterion for reporting potential severe accident sequences is consistent with that used for internal event IPEs. The steps in the progressive screening approach represent a series of analyses

in increasing levels of detail, effort, and resolution. However, the licensee may choose to bypass one or more steps so long as it identifies the vulnerabilities or demonstrates that they are insignificant. The screening approach consists of the following steps:

All plants:

1. Review plant-specific hazard data and licensing bases.

2. Identify significant changes since the operating license was issued.

3. Determine if the plant and facilities design meets the 1975 Standard Review Plan (SRP) criteria.

If the 1975 SRP criteria are not satisfied, or if it is known *a priori* that they will not be satisfied, one or more of the following approaches should be taken to further evaluate the situation.

Optional:

4. Determine if the hazard frequency is acceptably low.

5. Perform a bounding analysis.

6. Perform a probabilistic risk assessment (PRA).

Alternative Methods

The staff recognizes that other methods capable of identifying plant-specific vulnerabilities to severe accidents may be acceptable. A licensee may request that the staff review any other systematic examination method to determine if it is acceptable for IPEEE purposes.

Coordination with Ongoing Programs

Guidance is provided on coordinating the IPEEE process with ongoing programs. The first level of coordination is among the major elements related to the implementation of the Severe Accident Policy, that is, coordination among the IPEEE, the internal events IPE, containment performance improvements, and accident management. The second level of coordination is among the major elements of the IPEEE, that is, seismic events, fires, and high winds, floods, and others. The third level of coordination is within each major element of the IPEEE.

Programs subsumed into the IPEEE include the external event aspect of Unresolved Safety Issue (USI) A-45 (decay heat removal), Generic Issue (GI)-131 (in-core flux mapping system) and the Eastern U.S. Seismicity (Charleston earthquake) Issue. Programs that need to be coordinated with the IPEEE include USI A-46 (Seismic Equipment Qualification, which also covers the seismic spatial systems interaction of USI A-17 and the concern of USI A–40 for the seismic capability of large safety-related above-ground tanks), and GI–57 (Effects of Fire Protection System Actuation on Safety Related Equipment).

Peer Review

The licensee should conduct a peer review by individuals who are not associated with the initial evaluation to ensure the accuracy of the documentation package and to validate both the IPEEE process and its results.

ACKNOWLEDGEMENTS

This document represents the staff position on the Individual Plant Examination for severe accident vulnerabilities due to external events (IPEEE). Representatives of both the Office of Nuclear Regulatory Research and the Office of Nuclear Reactor Regulation were active contributors to the development of this guidance document; they are named below. In addition, significant input was received from consultants and contractors to the NRC, who are also named below. Edward Hill, of NRC, provided technical editing.

NRC

Charles E. Ader	Guy A. Arlotto	William D. Beckner
Goutam Bagchi	Demetrios L. Basdekas	Kazimieras M. Campe
T. Y. Chang	John T. Chen	Thomas M. Cheng
Nilesh C. Chokshi	Adel A. El-Bassioni	John H. Flack
R. Wayne Houston	David C. Jeng	Glenn B. Kelly
Roger M. Kenneally	Thomas L. King	P. T. Kuo
Conrad McCracken	Warren Minners	Jocelyn A. Mitchell
Thomas E. Murley	Andrew J. Murphy	David P. Notley
Thomas M. Novak	Leon Reiter	James E. Richardson
Robert Rothman	Lawrence C. Shao	Brian Sheron
Jack Strosnider	Rex G. Wescott	

Consultants

R. J. Budnitz	Future Resources Associates, Inc.
G. E. Cummings	Lawrence Livermore National Laboratory
R. P. Kennedy	Structural Mechanics Consulting, Inc.
M. K. Ravindra	EQE Engineering, Inc.

Contractors and Subcontractors

P. Amico	Science Application International
D. L. Bernreuter	Lawrence Livermore National Laboratory
M. P. Bohn	Sandia National Laboratories
R. J. Budnitz	Future Resources Associates, Inc.
D. H. Chung	Lawrence Livermore National Laboratory
B. C. Davis	Lawrence Livermore National Laboratory
G. S. Hardy	EQE Engineering, Inc.
J. R. McDonald	Texas Tech University
R. C. Murray	Lawrence Livermore National Laboratory
A. M. Nafday	EQE Engineering, Inc.
P. G. Prassinos	Lawrence Livermore National Laboratory
M. K. Ravindra	EQE Engineering, Inc.
J. Savy	Lawrence Livermore National Laboratory

1 INTRODUCTION

1.1 Background

On August 8, 1985, the Nuclear Regulatory Commission issued a policy statement on severe accidents (NRC, 1985). The Commission concluded, based on available information, that existing plants pose no undue risk to the public health and safety and that there is no present basis for immediate action on any regulatory requirements for these plants. However, the Commission recognizes, based on NRC and industry experience with plant-specific probabilistic risk assessments (PRAs), that systematic examinations are beneficial in identifying plant-specific vulnerabilities to severe accidents that could be fixed with low-cost improvements. As part of the implementation of the policy statement, the Commission issued Generic Letter 88-20 (NRC, 1988 and 1989), requesting that each licensee conduct an individual plant examination (IPE) for internally initiated events.

Risk assessments indicate that the risk from external events could be a significant contributor to the core damage in some instances. However, licensees were requested to proceed with the examinations only for internally initiated events (including internal flooding) in Generic Letter 88-20. Examination of severe accident vulnerabilities due to externally initiated events (IPEEE) is proceeding separately and on a later schedule to allow the staff to carry out additional work (SECY-88-147) to (1) identify which external hazards need a systematic examination, (2) identify examination methods and develop procedural guidance, and (3) coordinate the IPEEE with other ongoing NRC programs that deal with various aspects of external event evaluations to ensure that there is no duplication of industry efforts.

To accomplish these objectives, the staff established the External Events Steering Group (EESG) in December 1987 to make recommendations regarding the scope, methods, and coordination of the IPEEE (Beckjord, 1987, 1988). Specifically, the EESG is responsible for developing broad guidance for dealing with (1) external events on a generic basis both organizationally and technologically and (2) the implementation of the severe accident policy with respect to external events. The EESG established three technical subcommittees dealing with earthquakes (seismic events), internal fires, and high winds, floods, and "other" external events. The subcommittees were chartered to define the scope of the external events examination, identify acceptable examination methodologies, and coordinate ongoing issues and activities (for example, Unresolved Safety Issues and Generic Issues).

The EESG completed its task in May 1990. Based on the EESG recommendations, the staff prepared this report to provide detailed guidance on the conduct of the IPEEE and on the structure and content of the IPEEE submittal. It issued a draft of this report for public comment in July 1990. It conducted a workshop in September 1990 to explain the IPEEE process and to obtain specific comments and questions on draft Supplement 4 to Generic Letter 88-20 and the draft of this guidance document. In addition to the comments raised during the workshop, the staff received written comments from 16 organizations. This final report includes changes resulting from the resolution of these comments.

1.2 IPEEE Objectives

The objectives of the IPEEE, which are similar to the objectives of the internal event IPE, are for each licensee:

1. to develop an appreciation of severe accident behavior,

2. to understand the most likely severe accident sequences that could occur at the licensee's plant under full power operating conditions,

3. to gain a qualitative understanding of the overall likelihood of core damage and fission product releases, and

4. if necessary, to reduce the overall likelihood of core damage and radioactive material releases by modifying, where appropriate, hardware and procedures that would help prevent or mitigate severe accidents.

However, the staff recognized at the outset that the external initiators could not necessarily be treated in exactly the same way as internal initiators in the implementation of the Severe Accident Policy because the sources and treatment of uncertainties in estimates of core damage frequencies for external and internal events can be quite different. In addition, some methods endorsed by the staff for evaluating external hazards and identifying vulnerabilities do not produce estimates of core damage frequency. For example, seismic margins methods produce estimates of seismic hazard levels of high confidence-low probability of failure (HCLPF) for a plant rather than estimates of core damage probability.

Therefore, the staff determined that an explicit estimate of core damage frequency was not needed to meet the intent of the Severe Accident Policy and would not be a condition of the IPEEE. Thus, Objective 3 above would be addressed only indirectly by some methods acceptable for use in the IPEEE. Nevertheless, the key objective of gaining an understanding of plant behavior through the examination process could be met.

1.3 Purpose of Document

The purpose of this document is to provide guidelines for conducting the IPEEE and on the structure and content of the IPEEE submittal. It is not the intent of NUREG–1407 to go beyond the information request contained in Supplement 4 to Generic Letter 88–20. The external events recommended for inclusion in the IPEEE are identified in Section 2. Acceptable methodologies for performing an IPEEE along with upgrades to reflect state-of-the-art improvements are identified in Sections 3 through 5. Section 3 addresses the seismic portion; Section 4 the internal fires portion; and Section 5 the high winds, floods, and other portion of the IPEEE. Coordination between the IPEEE and the internal events IPE, other external events, and ongoing programs within each external event are provided in Section 6. A discussion of the peer review is provided in Section 7. A summary of documentation and reporting guidelines is provided in Section 8. The staff's responses to public comments and questions raised during the IPEEE workshop held in September 1990 and the written comments received soon afterward are given in Appendix D.

2 EVENTS EVALUATED FOR INCLUSION IN THE IPEEE

The external events considered, consistent with past probabilistic risk assessments (PRAs), are those events whose cause is external to all systems used in normal operation and emergency operation situations. Internal fire and internal flood are external to the "system" and therefore have been considered as external events in past PRAs. However, internal floods are being considered in the internal events IPE process (NRC, 1988).

In supporting the implementation of the Severe Accident Policy, a study of the risk of core damage to nuclear power plants in the United States due to externally initiated events was performed. The objective was to determine which external initiators have the potential of initiating an accident that may lead to severe reactor core damage or large radioactive release to the environment. Seismically initiated events are investigated in NUREG/CR–5042, Suppl. 1; internal fires, high winds/tornadoes, external floods, and transportation accidents are investigated in NUREG/CR–5042; "other external events" are investigated in NUREG/CR–5042, Suppl. 2. The "other external events" covered are nearby industrial/military facility accidents, on-site hazardous material storage accidents, severe temperature transients, severe weather storms, lightning strikes, external fires, extraterrestrial activity, volcanic activity, earth movement, and abrasive windstorms.

Some external events may not pose a significant threat of a severe accident to all plants, some events may have been considered in the plant's design to a sufficient degree, and some events may have been or will be reviewed under ongoing programs at some plants. The staff's evaluation and recommendations are contained in the following sections.

2.1 Seismic Events

The following are based upon an examination of current seismic design criteria, previous and ongoing seismic issues and programs, and seismic PRAs:

1. Mean seismic core damage frequencies calculated from past PRAs (NUREG-1150, NUREG/CR-5042, Suppl. 1) have been found to be in the range of 10^{-4} to 10^{-6} per year. Identified vulnerabilities are plant specific and include yard tanks, electrical equipment, diesel peripherals, structural failures, and equipment anchorages.

2. New data such as the occurrences of larger than anticipated earthquakes and the development of new hypotheses indicate that the plant-specific seismic hazard may be quite different from that envisioned at the time of licensing and make it difficult to rule out seismic events on the basis of initiating event frequencies.

3. Based primarily on their vintage, the current population of plants exhibit various levels of seismic design requirements and margin. Some of the very early plants have been backfitted under the Systematic Evaluation Program to ensure certain margins for safe shutdown using criteria different from current licensing criteria.

4. There have been modifications to plants since their original designs; for instance, the reduction of snubbers at some plants. These changes, in part, have relied on existing conservatism or risk-based arguments (e.g., LOCA + SSE combinations). The systematic examination of plants by the IPE and IPEEE will give an integrated picture of plants as they exist. It will also allow an integrated evaluation of the effects of individual changes made to plants over time.

5. There are unresolved safety issues and generic issues (e.g., USI A–45, USI A–46) that are in various stages of implementation. The IPE/IPEEE provides a convenient as well as meaningful framework for addressing many of these issues.

6. PRAs and seismic margins evaluations have resulted in cost-effective plant-specific improvements.

Therefore, the seismic external hazard should be included in the IPEEE.

2.2 Internal Fires

Based upon the examination of past fire PRAs, the contribution of internal fires to the probability of core damage may be significant and is very plant specific (NUREG/CR–5042). However, the numerical results always contain large uncertainties. The fire risk scoping study (NUREG/CR–5088) further confirms the following:

1. The overall fire-induced core damage frequency for the four plants studied (Seabrook, Oconee, Limerick, and Indian Point) increased from the original PRA studies even though, for certain fire scenarios, there was a net decrease. For all plants reviewed, fire continues to represent a dominant risk contributor.

2. Most initiating event frequencies were increased based on a much more complete data base available on fire occurrences in nuclear power plants. Under currently applied risk assessment methodologies, this increase in initiating event frequency alone

results in a direct increase in overall fire-induced core damage frequency with all other factors remaining constant.

3. Use of an expanded data base on historical fire suppression times for nuclear power plants resulted in a suppression probability distribution with a lower probability of suppression within a given time than that assumed in the original risk assessments. Under current methodologies, this again results in an increase in fire-initiated core damage frequency with all other factors remaining constant.

4. Updated information on the ignition and damage thresholds of cable insulation in some cases resulted in lower thermal damage limits. In some cases, no change in damage limits was required. A decrease in the assumed thermal damage limits would, in general, be expected to lead to increased estimates of fire-initiated core damage frequency.

5. Plant modifications made as a result of Appendix R requirements reduced the core damage frequency at Indian Point and Limerick for the requantified areas by factors of ten and three, respectively. For Seabrook, the identified Appendix R plant modifications did not affect the requantified core damage scenarios for internal fires. The Oconee PRA had already incorporated Appendix R modifications and no modifications subsequent to its performance were identified. Hence no Appendix R impact could be identified for either Seabrook or Oconee.

6. A number of issues that were not addressed in the past fire PRAs (effectiveness of fire brigade, effectiveness of fire barrier, seismic/fire interactions, control system interactions, and effects of fire suppressants on safety equipment) could increase the vulnerability to fire.

Therefore, based on the above studies, the internal fire hazard should be included in the IPEEE.

2.3 High Winds and Tornadoes

For plants designed against NRC's current criteria, these events pose no significant threat of a severe accident because the current design criteria for wind are dominated by tornadoes having an annual frequency of exceedance of about 10^{-7}. However, older plants and some modern plants having facilities not designed against these criteria need a systematic examination to identify plant-specific vulnerabilities (NUREG/CR–5042).

2.4 External Floods

For plants designed against current criteria as described in Regulatory Guide 1.59 and applicable Standard Re-

view Plan sections, particularly Section 2.4, floods pose no significant threat of a severe accident because the exceedance frequency of the design basis flood, excluding floods due to failure of upstream dams, is judged to be less than 10^{-5} per year (Chery, 1985), and the conditional core damage frequency for a design basis flood is judged to be less than 10^{-1}. Thus core damage frequencies are estimated to be less than 10^{-6} per year for a plant designed against NRC's current criteria. However, the latest probable maximum precipitation (PMP) criteria published by the National Weather Service (NWS) call for higher rainfall intensities over shorter time intervals and smaller areas than have previously been considered; this could result in higher site flooding levels and greater roof ponding loads than have been used in previous design bases (GI 103). Licensees are requested to assess the effects of applying these new criteria to their plants in terms of onsite flooding and roof ponding. Also, some older plants may have higher potential risk and need systematic examinations for plant-specific vulnerabilities.

2.5 Transportation and Nearby Facility Accidents

These events consist of accidents related to transportation and accidents at industrial and military facilities. Plants designed against NRC's current criteria (NUREG/CR–5042) should have no significant vulnerability to severe accidents from these events because the initiators considered in the design should have a recurrence frequency less than 10^{-6} or have been shown through a bounding analysis not to affect the plant. However, changes may have occurred since the original design and there may be exceptions that need some systematic examination. Also, some older plants may not meet the NRC's current criteria and need systematic examinations for plant-specific vulnerabilities.

2.6 Lightning

Lightning has been experienced at many nuclear power plants in the United States (NUREG/CR–5042, Suppl. 2; AEOD, 1986; ACRS, 1989). The impact of lightning on plant operation and the vulnerability of plants to a severe accident due to lightning has been examined. The major conclusion is that the primary impact of lightning on nuclear power plants is the loss of offsite power. The loss of offsite power is included as part of the internal events IPE, and examination for vulnerabilities due to this aspect of lightning is therefore already included in the IPE process. The staff has concluded that, in general, other effects of lightning on nuclear power plants are insignificant. However, further examination of lightning effects may be warranted for certain sites where, based on past operating experience, lightning strikes are likely to cause more than just loss of offsite power; for example, they may also affect safety-related instrumentation and control systems.

Based on an examination of historical data on lightning, as well as knowledge of plant systems, the staff concludes the following:

1. Lightning has typically caused partial or complete loss of offsite power, which is the main impact of lightning and which is already being examined as part of the internal events IPE.

2. Lightning is much less likely to affect the onsite power system.

3. Lightning has affected safety-related equipment and has caused reactor trips, but these events have not been significant in terms of impact on the plant.

4. Safety systems (e.g., diesel generators, electrically powered pumps) are not normally in operation. Thus, while control systems may be damaged, the safety systems are less susceptible to damage and may be manually activated.

5. Redundancy of safety systems and the capability for recovery of systems (replacing fuses or resetting breakers) further reduce the likelihood that the low frequency of lightning damage events will result in a severe accident.

The staff has judged that the probability of a severe accident caused by lightning (other than one due to loss of offsite power) is relatively low and further consideration of lightning effects should be performed only for plant sites where lightning strikes are likely to cause more than just loss of offsite power or a scram.

2.7 Severe Temperature Transients (Extreme Heat, Extreme Cold)

Severe temperature transients may affect nuclear power plants in the United States (NUREG-1032). However, the effects are usually limited to reducing the capacity of the ultimate heat sink and loss of offsite power (NUREG/CR-5042, Suppl. 2). The capacity reduction of the ultimate heat sink would be a slow process that allows plant operators sufficient time to take proper actions such as reducing power output level or achieving safe shutdown, if necessary, and maintaining the plant in a safe shutdown condition. The other potential impact on the plant, loss of offsite power, will be considered within the realm of the station blackout rule (NRC, 1988b) and the internal event IPE. Therefore, the temperature transients need not be addressed in the IPEEE.

2.8 Severe Weather Storms

Severe weather storms (ice storm, hailstorm, snowstorm, dust storm, sandstorm) accompanied by strong winds have caused several complete and partial losses of offsite power (NUREG/CR-5042, Suppl. 2). The potential effect of loss of offsite power and station blackout will be addressed in the internal event IPE; thus severe weather storm evaluations need not be repeated in the IPEEE.

2.9 External Fires (Forest Fires, Grass Fires)

These are fires occurring outside the plant site boundary. Potential effects on the plant could be loss of offsite power and forced isolation of the plant ventilation and possible control room evacuation. Usually, external fires are unable to spread onsite because of site clearing during the construction stage (NUREG/CR-5042, Suppl. 2). However, there has been one instance during which a nearby forest fire caused a partial loss of offsite power. The effect of loss of offsite power will be addressed in the internal events IPE and need not be repeated in the IPEEE. The other effects have been evaluated during operating license (OL) review against sufficiently conservative criteria; thus they do not need to be reassessed in the IPEEE.

2.10 Extraterrestrial Activity (Meteorite Strikes, Satellite Falls)

Extraterrestrial activity is considered to be natural satellites such as meteors or artificial satellites that enter the earth's atmosphere from space. Because the probability of a meteorite strike is very small (less than 10^{-9}) (NUREG/CR-5042, Suppl. 2), it can be dismissed on the basis of its low initiating event frequency.

2.11 Volcanic Activity

Most nuclear power plant sites are too far away from active volcanoes to expect any effect at the plant, so most licensees need not consider the volcanic effects. However, those sites in the vicinity of active volcanoes should assess volcanic activities (NUREG/CR-5042, Suppl. 2) as part of the IPEEE process.

2.12 Summary

In summary, based on the above evaluation, five events need to be included by all licensees in the IPEEE: seismic events, internal fires, high winds, floods, and transportation and nearby facility accidents. All licensees should confirm, however, that no plant-unique external events known to the licensee today with potential severe accident vulnerability are being excluded from the IPEEE.

3 ACCEPTABLE METHODOLOGIES FOR PERFORMING THE SEISMIC IPEEE

For the purpose of performing an IPEEE, two methodologies are considered acceptable to identify potential seismic vulnerabilities at nuclear power plants. The first is a seismic probabilistic risk assessment (PRA) (NUREG/CR–2300; NUREG/CR–2815, Vol. 2; NUREG/CR–4840). In addition, an evaluation of the reliability and usefulness of results and insights obtained from external event PRA methodologies is contained in NUREG/CR–5477. The second method is one of the seismic margins methodologies (SMM) described in NUREG/CR–4334, NUREG/CR–4482, and NUREG/CR–5076, EPRI NP–6041 or the reduced SMM described later in this section.

In meeting the objectives of the IPEEE, the examination should focus on qualitative insights from the systematic plant examination rather than only on absolute core damage frequency estimates. Guidance for performing the seismic IPEEE using a PRA or margins methodology is provided below.

3.1 Seismic PRA

This discussion deals with the use of PRA techniques in the seismic IPEEE. The PRA should be at least a Level 1 plus containment performance analysis. The basic elements are (1) hazard analysis, (2) plant system and structure response analysis, (3) evaluation of component fragilities and failure modes, (4) plant system and sequence analysis, and (5) containment and containment system analysis including source terms, to identify unique seismic sequences or vulnerabilities different from the internal event analysis. Specific guidance and enhancements are provided for licensees performing a new PRA or updating an existing seismic PRA.

3.1.1 New Seismic PRA Analysis

3.1.1.1 General Considerations

Licensees choosing to do a seismic PRA built on an internal events PRA should be aware of important considerations that, if incorporated in the planning of the internal events PRA, will minimize their resource expenditure and speed the staff reviews. For example, (1) a well-organized walkdown team and a properly planned walkdown will enable many issues to be addressed at the same time; (2) the peer review team should consider the need to review both internal and external event analyses; (3) fault tree analysts for internal events should be aware of spatial interactions (including internal flooding effects), failure of passive components such as structures

and supports, and common-cause effects (the culling or pruning of trees should be done with these considerations in mind); and (4) internal event models should be developed knowing that, in the seismic analysis, the fragilities of a component are sensitive to elevation. Also, a component and its peripheral equipment may have different fragilities. Additional discussion of this subject can be found in NUREG/CR–4840.

PRA calculations that account for all uncertainties are clearly acceptable. However, the staff believes that, for the seismic IPEEE, it is not necessary to carry out complete uncertainty quantifications defining a distribution of core-damage frequencies in order to identify vulnerabilities. Mean point estimation using a single hazard curve (rather than a family of hazard curves) and a single fragility curve (rather than a family of fragility curves) for each component is sufficient to get insights into potential seismic vulnerabilities. To highlight the most pertinent results/insights from the seismic portion of the IPEEE, mean point estimates using hazard curves described in NUREG/CR–5250 and EPRI NP–6395D should be obtained. Further discussion on the use of hazard curves is contained in Section 3.1.1.2.

The above point estimation approach is valid only because of the IPEEE objective: to identify dominant sequences and components and where possible rank them. (This point estimate should not be confused with a "Phase I" type PRA analysis where point estimate calculations are used only to define scopes for more detailed Phase II and Phase III studies). Fragilities used in this point estimate, where possible, should be plant specific and rigorous to be able to identify dominant components and rank them. Correlations and other aspects of analysis should be treated so that, when a mean seismic hazard curve is used with the mean plant fragility curve, the resulting core damage estimate approximately represents the mean estimate that would be derived from the full uncertainty analysis.

The recommendation of performing point estimation type calculations is made primarily to highlight insights needed for the severe accident behavior perspective. This should not be construed as de-emphasizing or ignoring uncertainties. Analysts are encouraged to make careful study of the origins of the possible uncertainties, including those that are hardest to quantify. Many of the insights obtained from a PRA analysis are obtained by trying to gain a better understanding of the uncertainties. Consideration of uncertainties may affect how the results of a PRA are implemented in plant changes.

3.1.1.2 Hazard Selection

For the United States east of the Rocky Mountains, two highly sophisticated seismic hazard studies were conducted by Lawrence Livermore National Laboratory (LLNL) (NUREG/CR–5250) and the Electric Power Research Institute (EPRI) (EPRI NP–6395–D). For many sites, these studies yield significant differences at the low-probability and high-level ground motions. The initial PRAs carried out using these estimates (Surry and Peach Bottom in NUREG–1150) indicate that, despite large differences in absolute numerical estimates, the identification, ranking, and relative contributions of the dominant seismic sequences are virtually the same for both LLNL and EPRI hazard estimates. This equivalence is apparently due to the fact that the slopes of the seismic hazard curves are not significantly different over those ground motion levels that, in conjunction with the fragilities, control the relative distribution of seismically induced core damage frequencies. Although these results are very encouraging, there is no guarantee that this will be true for all sites in the Central and Eastern United States.

While a full seismic hazard uncertainty analysis is not necessary in performing a seismic PRA for the IPEEE, the staff prefers that mean (arithmetic) hazard estimates from both the LLNL and the EPRI studies should be used to obtain two different point (mean) estimates. If a licensee elects to perform only one analysis, it should use the higher of the two mean (arithmetic) hazard estimates.

The use of both the LLNL and EPRI mean hazard curves has another advantage in that the extent of uncertainty will become obvious and the emphasis on the bottom line numbers is reduced. The use of both of these estimates (LLNL and EPRI) will serve to identify differences, if any, in the delineation of dominant seismic sequences (minor variations in contributions and rankings are anticipated). Such differences would have to be addressed by the licensee in its identification and listing of vulnerabilities.

For plants in the Western United States, for which there are no counterparts to the LLNL and EPRI studies, a licensee should conduct its own study to define the mean hazard estimate for use in the IPEEE. The licensee should also provide reasonable assurance that any significant uncertainty in those elements of hazard (for example, slope) that control the identification, ranking, and relative contribution of seismic contributors to core damage is addressed in sensitivity studies. As in the Central and Eastern United States, the identification and listing of vulnerabilities should take this uncertainty into account.

Most seismic PRAs use peak ground acceleration as the hazard parameter. If this is done, spectral shapes that are consistent with current estimates of ground motion should be used. In the Central and Eastern United States, current spectral estimates can be found in the LLNL and EPRI studies. Since similar spectral shapes are obtained from LLNL and EPRI hazard studies, separate analyses using both spectral shapes are not needed. Median spectral shapes of 10,000 year return period provided in NUREG/CR–5250 along with variability estimates are recommended for use in the analyses. Other site-specific spectral shape estimates may be proposed (that is, derived from a suite of appropriate recorded earthquakes). For the Western United States, site-specific spectral shape should be established and used. Since only one spectral shape is used for both hazard analyses, two separate plant response and fragility analyses are not needed.

If an upper bound cutoff to ground motion at less than 1.5 g peak ground acceleration is assumed, sensitivity studies should be conducted to determine whether the use of this cutoff affects the delineation and ranking of seismic sequences.

3.1.1.3 Fragility Estimation

The following guidance on fragility estimation is included to clarify the use of fragility in the context of the "point estimation" approach discussed above. Details and methods for fragility and high confidence of low probability of failure (HCLPF) calculations are discussed in a number of references, for example, NUREG/CR–2300, NUREG/CR–4334, NUREG/CR–4482, NUREG/CR–5076, NUREG/CR–4659, Vols. 1–3, EPRI–NP 6041, and NUREG/CR–5270. It is recognized that large uncertainties exist in fragilities estimation (NUREG/CR–5270). A perspective on how this uncertainty affects the results of analysis (numerical and other insights, for example, dominant sequences and components) should be maintained.

Consistent with the point estimation approach, one can use a single mean component fragility curve for each component and hence for sequence-level and plant-level assessments. This mean curve is defined by the median capacity, $\bar{a}$, and composite uncertainty, β_c, where $\beta_c^2 = \beta_r^2 + \beta_u^2$, when β_r and β_u are estimated separately. β_r and β_u represent random uncertainty and modeling uncertainty, respectively. It is also acceptable to use a family of fragility curves instead of a single curve.

When a single mean fragility curve is available, HCLPF capacity for a component (sequence, or plant) can be approximated by -2.3 β_c below the median (i.e., 1% composite probability of failure is essentially equivalent to 95% confidence of less than 5% probability of failure). While developing sequence-level and plant-level fragilities, the licensee should retain the ability to report HCLPFs with and without nonseismic failures and human actions.

3.1.1.4 Seismic PRA Methodology Enhancements

Review of past seismic PRAs indicates that certain areas have been treated either inconsistently or not at all. Therefore, the following areas should be included:

1. *Plant Walkdowns.* Walkdowns are performed to find as-designed, as-built, and as-operated seismic weaknesses in plants. Each licensee should perform a walkdown consistent with the intent of the guidelines described in Sections 5 and 8 and Appendices D and I of the EPRI Seismic Margins Methodology (EPRI NP-6041).

2. *Relay Chatter.* Relays, in this context, include components such as electric relays, contactors, and switches that are prone to chatter. Additional guidance is given in NUREG/CR-5499. The scope of the relay chatter evaluation should be consistent with the site's seismic margin review level earthquake classification (full scope or focused scope) as identified in Tables 3.1 and 3.2 and discussed in Sections 3.2.4.2 and 3.2.5.3. That is, a more complete evaluation is to be performed for sites in the full-scope category than for sites in the focused-scope category. It is anticipated that chatter and recovery actions will be modeled as necessary. The focused-scope evaluation can be limited to a review of low seismic ruggedness relays for plants that are not included in the USI A-46 program.

   The examination of the relay chatter effects (for example, the Hatch margin evaluation) has resulted in large resource expenditures in terms of staff-hours. Therefore, careful planning and use of generic insights, if they are applicable to the plant, are necessary. Additional guidance on this topic is also included in Section 3.2.4.2.

3. *Liquefaction.* The potential for soil liquefaction and associated effects on the plant need to be examined for some sites because of specific site conditions. The impact on plant operation should be assessed from the point of view of both potential for and consequences of liquefaction. Procedures for assessing soil liquefaction are described in EPRI NP-6041.

4. *HCLPF Calculations (Optional).* Licensees can report plant-level, sequence-level, and component-level HCLPF values and use this information to support decisions related to the identification and listing of vulnerabilities. In several PRAs (for example, Millstone 3 and Diablo Canyon), HCLPF estimates are reported along with other PRA results. These PRAs can be used as guidance for deriving HCLPF values from fragilities. HCLPF values are to be reported both with and without the effects of nonseismic failures and human actions. If a licensee

does not supply HCLPF calculations, the staff will calculate the HCLPF values based on information provided in the IPEEE submittal and will use them in the evaluation of the submittal. Note that plant-level, sequence-level, and component fragilities are to be documented.

3.1.1.5 Containment Performance

The primary purpose of the containment performance evaluation is to identify sequences and vulnerabilities that involve containment, containment functions, and containment systems (e.g., igniters and suppression pools) seismic failure modes or timing that are significantly different from those found in the IPE internal events evaluation. Additional guidance is presented in Section 3.2.6.

3.1.2 Use of an Existing PRA

The use of an existing seismic PRA to address the seismic IPEEE is acceptable provided the PRA reflects the current as-built and as-operated condition of the plant and some of the deficiencies of past PRAs discussed below are adequately addressed.

1. *Hazard Selection.* For PRAs at sites east of the Rocky Mountains that did not use the LLNL and EPRI mean hazard estimates, sensitivity studies should be conducted to determine if the use of these results would affect the delineation or ranking of seismic sequences. For PRAs in the Western United States, the sensitivity studies should be carried out to determine the effect of uncertainty in hazard on the delineation and ranking of seismic sequences.

2. *Walkdowns.* Since a walkdown is considered to be one of the most important ingredients of the seismic IPEEE, a supplementary walkdown in conformance with the intent of the procedures described in Sections 5 and 8 and Appendices D and I of the EPRI margin methodology (EPRI NP-6041) should be performed. It may be necessary to amplify the earlier analysis based on the walkdown outcome. These results should be reported.

3. *Relay Chatter.* Relay chatter effects either have not been considered or were assumed fully recoverable in past PRAs. Relays, in this context, include components such as electric relays, contactors, and switches that are prone to chatter. Licensees should analyze the effect of relay chatter or determine that the type of relays used in the safety systems are not subject to relay chatter. The scope of the review should be consistent with the site's seismic margins review level earthquake classification (full scope or focused scope) as identified in Tables 3.1 and 3.2 and discussed in Sections 3.2.4.2 and 3.2.5.3. Additional guidance is provided in NUREG/CR-5499. Results of this effort that lead to a PRA revision or plant fixes should be reported.

Table 3.1 Review Level Earthquake—Plant Sites East of the Rocky Mountains

Reduced Scope

Big Rock Point	Duane Arnold*	South Texas	Turkey Pt.
Comanche Peak	Grand Gulf	St. Lucie	Waterford
Crystal River	River Bend		

0.3g Focused Scope

Arkansas #2	Davis-Besse	Limerick	Salem
Beaver Valley	Dresden	McGuire	Shoreham
Bellefonte	Farley	Millstone	Summer*
Braidwood	Fermi	Monticello	Surry
Browns Ferry	Fitzpatrick	Nine Mile Pt.	Susquehanna
Brunswick	Fort Calhoun	North Anna*	Three Mile Island
Byron	Ginna	Oyster Creek	Vermont Yankee
Callaway	Haddam Neck	Palisades	Vogtle
Calvert Cliffs	Harris	Peach Bottom	Watts Bar
Catawba*	Hatch	Perry	Wolf Creek
Clinton	Hope Creek	Point Beach	Zion
Cook	Kewaunee	Prairie Island	
Cooper	LaSalle	Quad Cities	

0.3g Full Scope

Arkansas #1	Maine Yankee	Robinson	Yankee Rowe
Indian Point	Oconee*	Sequoyah	

Committed to Perform a Seismic PRA

Pilgrim**	Seabrook**

Notes:
 *Special attention to shallow soil conditions is appropriate for these locations (see Section 3.2.2).
 **Relay chatter evaluation should be similar to a full-scope review.

Table 3.2 Review Level Earthquake—Western United States Plant Sites

0.5g*

Trojan	Rancho Seco
Washington Nuclear	Palo Verde

Seismic Margin Methods Do Not Apply to the Following Sites:

Diablo Canyon	San Onofre

Notes:
 *Indicates a site in the Western United States whose default bin is 0.5g, unless the licensee can demonstrate that the site hazard is similar to that at sites east of the Rocky Mountains that are found in the 0.3g bin.
 Changes in the review level earthquake from 0.5g to 0.3g should be approved before doing significant analysis.

4. *Nonseismic Failures and Human Actions.* In several seismic PRAs, nonseismic failures (e.g., failures of the auxiliary feedwater system and failure of feed and bleed mode of core cooling, battery depletion, power-operated relief valve failures) and human actions (e.g., delays or failures in performing specified actions or operator misdiagnoses a situation and takes an improper action that is not related to the actual, current plant situation) have been important contributors to seismically induced core damage frequencies or risk indices. Unless nonseismic failures are considered, improper decisions may be made regarding plant modifications or procedural changes.

The licensee has the option to expand its PRA or demonstrate that the exclusion of nonseismic failures will not significantly alter the PRA results or insights. The scope of nonseismic failures and human interactions that might affect seismic sequences should be defined by the licensee based on the internal events analyses.

5. *Liquefaction.* The potential for soil liquefaction and associated effects on the plant need to be examined for some sites because of specific site conditions. The impact on plant operation should be assessed from the point of view of both potential for and consequences of liquefaction. Procedures for assessing soil liquefaction are described in EPRI NP-6041.

6. *Containment Performance.* Licensees should ensure that the performance of containment and containment systems are addressed. Section 3.2.6 contains guidance.

7. *HCLPF Calculations (Optional).* Licensees can extract and report plant-level, dominant-sequence-level, and dominant-component-level HCLPFs both with and without the effects of nonseismic failure and human actions and use this information to support decisions related to the identification and listing of vulnerabilities. If the licensee does not supply HCLPF calculations, the staff will calculate HCLPF values based on information provided in the IPEEE submittal and will use them in the evaluation of the submittal.

3.2 Seismic Margin Methodologies

This discussion deals with the use of the seismic margin methodology in the seismic IPEEE. Specifically, guidance and enhancements are provided for a licensee using either the NRC or EPRI margins methodology.

3.2.1 General Considerations

The seismic margin methodology is considered acceptable for addressing seismic concerns in the severe accident policy implementation. Two methodologies are currently available: one developed under NRC sponsorship and the other developed under EPRI sponsorship. The staff has determined that both methods (with the noted enhancements) will adequately address IPEEE objectives.

The two methods use different system analysis philosophies. The NRC method is based on an event/fault tree approach to delineate accident sequences. For example, for PWRs, two safety functions are considered to be most important to plant seismic safety: reactor subcriticality and early emergency core cooling. If these functions are ensured for a given earthquake, there is high confidence that core damage would not occur at that level. The EPRI methodology is based on a systems "success path" approach. This approach defines and evaluates the capacity of those components required to bring the plant to a stable condition (either hot or cold shutdown) and maintain that condition for at least 72 hours. Several possible success paths may exist. Both the NRC and the EPRI methods were used in the trial application at the Hatch plant. Application of the NRC method was limited to a systems review. Insights gained from the use of these two methods and the differences between them are discussed in reports by Davis (1990), Orvis et al. (1990), and Shao et al. (1990).

Each licensee should examine its plant critically to ensure that the generic insights used in margin methodology development to identify critical functions, systems, and success path logic are applicable to its plant. This is particularly vital for older plants where systems and functions may differ greatly from the plants considered in the development of the margins methodologies (NUREG/CR-4334, NUREG/CR-4482, NUREG/CR-5076, and EPRI NP-6041).

Based on written comments on draft of NUREG-1407 and public meetings, the staff has defined three categories of margin studies requiring varying levels of effort. The three categories are full-scope, focused-scope, and reduced-scope. The focused-scope category is new, while the full-scope (although not identified as such in draft NUREG-1407) and reduced-scope categories are retained. The primary purpose of this further subdivision is to reduce the level of effort for some plants. Licensees of plants with relatively higher hazard and a lower seismic design basis will conduct more detailed studies than will licensees of other plants (grouping of plants into various categories is discussed in Section 3.2.2 and Appendix A).

The examination scope in each category for NRC and EPRI methods is discussed in detail in Sections 3.2.4 and

3.2.5. An introductory discussion of the rationale and the general characterization of full-scope and focused-scope follows.

The major difference between the full scope and the focused scope is the scope of relay chatter evaluation. Based on detailed studies carried out at the Hatch and Diablo Canyon plants at considerable resource expenditure, it appears that the problems that could be caused by relay chatter at these plants were recoverable with existing procedures. However, there is a concern among the staff and industry consultants that such conclusions can not be considered generic without some additional plant reviews. Additionally, both the NRC-sponsored and the EPRI-sponsored relay tests indicate that relay performance is very sensitive to variables such as spring tension, orientation, and mounting. Tests further indicate that a significant number of relays may have capacities between SSE and RLE levels. USI A–46 evaluations are to be performed only at the SSE levels. Therefore, the staff is recommending that the full-scope plants evaluate relay chatter in a manner consistent with the approach suggested in EPRI NP–6041 or equivalent. Note that considerable efficiency can be achieved using lessons learned from the Hatch experience (Moore et al., 1990). For plants in the focused-scope category, a low level of effort is recommended. The lessons learned from the full-scope reviews may necessitate reexamining the relay chatter issues through the generic issues process (i.e., NUREG–0933).

Other differences between the full- and focused-scope relate to the level of effort for evaluating soil failure modes and the number of margin calculations (HCLPFs) to be reported (Reed et al. 1990; Rasin, 1990). The difference in the level of effort in these areas stems primarily from a perceived need for more accurately characterizing plant behavior and numerical results for plants in the full-scope category. It is also perceived that licensees of plants in the focused-scope category will be able to identify important vulnerabilities with more liberal use of fewer, approximate, and bounding-type analyses without adverse impact. However, the actual level of effort in these areas is very much site and plant dependent and should be determined on the basis of plant-specific considerations. For example, a plant in the full- scope category that is located on a rock site will not perform any soil failure evaluation, while a plant in the focused-scope category that is located at a coastal plain site may require more sophisticated investigations. In any case, discussions here do not preclude the use of properly substantiated judgments to define the scope and depth of an examination.

3.2.2 Review Level Earthquake and Associated Spectral Shapes

The seismic margins methodology was designed to demonstrate sufficient margin over SSE to ensure plant safety and to find any "weak links" that might limit the plant shutdown capability to safely withstand a seismic event bigger than SSE. The seismic margin method utilizes two review or screening levels geared to peak ground accelerations of 0.3g and 0.5g. It is the staff's judgement that the use of a 0.3g review level earthquake (RLE) for most of the nuclear power plant sites in the Central and Eastern United States (east of the Rocky Mountains) would serve to meet the objectives of the IPEEE. However, all sites east of the Rocky Mountains are not subject to the same level of earthquake hazard. For some sites where the seismic hazard is low, a reduced-scope margin approach centered on walkdowns is acceptable. For western sites other than California coastal sites, a 0.5g RLE should be used for the margin approach. The RLEs defined for U.S. sites, as well as sites that can perform a reduced-scope SMM, are presented in Tables 3.1 and 3.2. The seismic margin evaluations should utilize the NUREG/CR-0098 median rock or soil spectrum anchored at 0.3g or 0.5g depending on the g level and primary condition at the site. Further discussion on the review level earthquake is presented in Appendix A.

Plants in the 0.3g bin are further subdivided into full- and focused-scope categories, as discussed earlier. This categorization is based on consideration of hazard as well as the seismic design basis. Additional consideration is also given to the outlier plants resulting from the Eastern U.S. Seismicity (the Charleston Earthquake) Issue resolution. Additional details are given in Appendix A.

The ground motion should be considered at the surface in the free field. If secondary conditions such as shallow soil conditions are being considered, appropriate procedures should be used to determine the free-field motion in the vicinity of those affected structures and components, and the capacity evaluation of structures and components should take into account the effects of soil-structure interaction.

Because recent ground motion estimates, such as those included in the LLNL and EPRI hazard studies, indicate relatively higher ground motion at frequencies greater than 10 Hz than that shown in the NUREG/CR-0098 spectrum, the margin evaluation of only nonductile components (if *appropriate*)—for instance, relays—that are sensitive to high frequencies should be performed as discussed in Section 3.2.4.2. No plant-specific response analysis is anticipated to address concerns related to high-frequency ground motion. However, if a licensee decides to evaluate plant response for high-frequency ground motion, the response spectral shapes derived from the appropriate site-specific median uniform hazard response spectra (10,000-year return period) shown in NUREG/CR–5250 anchored at 0.3g or 0.5g should be used.

3.2.3 Reduced-Scope Margins Method

For sites where the seismic hazard is low, a reduced-scope seismic margins method emphasizing the walkdown is adequate. Well-conducted, detailed walkdowns have been demonstrated to be the most important tool for identifying seismic weak links whose correction is highly cost effective. Applicable sites are identified in Table 3.1.

The initial steps of the full-scope margin methodology up to and including the initial plant walkdown are performed regardless of method selected (NRC or EPRI). Basically, pertinent activities up to and including the initial plant walkdown need to be performed. These activities include gathering system information, classifying front-line systems and identifying front-line components, classifying support systems and identifying support system components, and identifying plant-unique features.

Further guidance on the differences between the reduced-scope and full-scope seismic margins methods, that is, elements preserved and elements eliminated are provided in Appendix B.

The evaluation should be documented in a walkdown team report and subjected to a peer review (see Section 7).

3.2.4 NRC Seismic Margins Methodology

The guidance in NUREG/CR–4334, NUREG/CR–4482, and NUREG/CR–5076 is supplemented by that in the following sections to (1) reflect the partitioning of the 0.3g screening criteria into the reduced-scope, 0.3g focused-scope, and 0.3g full-scope bins identified in Table 3.1 and (2) identify enhancements so that the method can be used for IPEEE implementation.

3.2.4.1 Walkdown

Reduced Scope

See Appendix B, Sections B.1 (1) through (4) for guidance.

Emphasis on walkdowns also applies to containment and containment systems (that is, containment functions required to prevent early failure, containment integrity, isolation, and prevention of bypass), USI A–45, and GI–131.

Focused Scope, Full Scope, and 0.5g

The walkdown should be performed and documented in accordance with the recommendations contained in EPRI NP–6041.

3.2.4.2 Relay Evaluation

Relays in this context include such components as electric relays, contactors, and switches that are prone to chatter.

The following paragraphs define the scope of the relay chatter evaluation for each of the three bins:

Reduced Scope

USI A–46 Plants—Completion of the USI A–46 review will satisfy the IPEEE intent.

Non A–46 Plants—No action is needed.

Focused Scope

USI A–46 Plants—Follow USI A–46 procedures. If low-seismic-ruggedness relays are discovered during the USI A–46 review, the relay review should be expanded to include relays outside the scope of USI A–46 but within the scope of the IPEEE.

Non A–46 Plants—Locate and evaluate low-seismic-ruggedness relays (bad actor list).

Full Scope and 0.5g (Including Western U.S. Plants)

USI A–46 Plants—Follow USI A–46 procedures for the review; review systems within the scope of the IPEEE, including those that are also within the scope of USI A–46, using appropriate margin (EPRI NP–6041) or USI A–46 procedures at the RLE.

Non A–46 Plants—Review the relays in all systems within the scope of the IPEEE, using appropriate margin (EPRI NP–6041) or USI A–46 procedures at the RLE.

The NRC method as originally developed did not address the relay chatter issue because information on the subject was lacking. Hardy et al. (1989), summarized the results of several efforts in this area and provided guidance to address this issue in an IPEEE context. Relay chatter analysis could be resource intensive, and careful planning and use of generic insights, if they are applicable to the plant, are desirable. Insights and recommendations based on the Hatch experience are available in Moore et al. (1990).

Attempts to address the concerns related to high-frequency ground motion by analysis is very likely to entail extensive efforts, including the development of new and much more complex building models that transmit and amplify high-frequency input and generate accurate and meaningful floor spectra at high frequencies. Estimates of high-frequency amplification in cabinets containing relays will also have to be developed. Rather than using analysis, the following approach is more suitable:

1. Prepare a list of relays that are known to have high-frequency sensitivity.

2. Screen relays that are known to have very high HCLPFs (that is, eliminate them from further consideration without performing specific response calculations).

3. Assume that the remaining relays will chatter at the review level earthquake.

4. Screen the remaining relays by showing either that the electrical circuity is insensitive to high-frequency chatter or that they can be recovered from changes of state and associated false alarms.

5. Finally, replace the remaining relays with relays that are not sensitive to high frequency (an alternative approach is to show that the remaining relays are rugged by conducting tests).

Although stated in the context of high-frequency ground motion, this approach can be used to address the relay chatter issue.

3.2.4.3 Soil Failures

Soil failure analyses include an evaluation for instability, settlement, and liquefaction.

Reduced Scope

No evaluation is necessary.

Focused Scope

EPRI NP-6041 contains guidance; a review based on appropriate design and construction records is considered adequate. A detailed analysis, as necessary, will be performed if soil failure is deemed significant.

Full Scope and 0.5g

EPRI NP-6041 contains guidance; it is anticipated that existing soil test data will be adequate. An evaluation of plant site conditions using state-of-the-art approaches should be performed if soil failure is deemed significant.

3.2.4.4 Screening Criteria (Use of Screening Tables)

The screening guidance given in the Generic Implementation Procedure for Seismic Verification of Nuclear Power Plant Equipment (GIP) may be used, provided a review is conducted at the appropriate RLE, caveats included in margins reports are observed, and limitations on the use of the generic equipment ruggedness spectrum

(GERS) are observed. Also, spatial interaction evaluation, such as assessing the effects of flooding, as noted in EPRI NP-6041, is retained.

Reduced Scope

Appendix B contains guidance.

0.3g Bin—Focused and Full Scope

The criteria in the <0.3g column of NUREG/CR-4334, Table 5.1, or EPRI NP-6041, Tables 2-3 and 2-4, should be used.

0.5g Bin

The criteria in the 0.3-0.5g column of NUREG/CR-4334, Table 5.1, or EPRI NP-6041, Tables 2-3 and 2-4, should be used.

3.2.4.5 Seismic Input

Reduced Scope

For the evaluation, the SSE ground response spectra and in-structure spectra should be used. New in-structure response spectra, if developed, should be mean plus one standard deviation to be consistent with the conservatism in the design input. Any differences between the Final Safety Analysis Report (FSAR) and new response spectra should be highlighted and discussed.

Focused-Scope, Full-Scope, and 0.5g

For the evaluation, the NUREG/CR-0098 median rock or soil spectrum (depending on primary condition at the site) anchored at the assigned review level earthquake should be used.

3.2.4.6 Evaluation of Outliers—HCLPF Calculations

Two approaches, fragility analysis (FA) and conservative deterministic failure margin (CDFM), for computing component and plant HCLPFs are acceptable. For the NRC margins method, if the licensee initially chose the CDFM method to calculate component and plant HCLPF values, it is possible to make plant HCLPF statements with and without the inclusion of nonseismic failures and human actions by developing complete fragilities for the few components that remain in the plant-level Boolean equations (optional).

As noted in EPRI NP-6041, use of the Generic Equipment Ruggedness Spectrum (GERS) to estimate HCLPFs should take into account the latest results from ongoing work on the reconciliation of GERS and HCLPF.

Reduced Scope

Outliers should be evaluated for the provisions in the GIP if the plant is also in the USI A-46 Program. For elements outside the USI A-46 scope (structures and piping) the requirements of the plant FSAR should be used in the evaluation. Since the evaluation is done at the design level, the outliers should be addressed in accordance with 10 CFR 50.72(b).

Focused Scope

The seismic capability evaluation engineers/seismic review team may use judgment to rank the unscreened structures and equipment from the lowest to the highest. The licensee should determine the number, scope, and type of HCLPF analyses with the aim of identifying vulnerabilities and ranking them. Reed et al. (1990) and Rasin (1990) suggest that HCLPF capacities should be calculated for the lowest one-third of the ranked components; the remaining components should be assigned a conservative HCLPF based on the highest calculated HCLPFs. The assumed and calculated HCLPFs should be reported.

Full Scope and 0.5g

HCLPFs for unscreened structures and components should be calculated as needed to accurately characterize plant HCLPFs and vulnerabilities and rank them.

3.2.4.7 Nonseismic Failures and Human Actions

These activities should be included; guidance on including nonseismic failures and human actions is provided in NUREG/CR–4826 (Maine Yankee evaluation) and in two draft reports by Budnitz (1987 and 1990).

3.2.5 EPRI Seismic Margins Methodology

The guidance provided in EPRI NP–6041 is supplemented by that in the following sections to (1) reflect the partitioning of the 0.3g screening criteria into the reduced-scope, 0.3g focused-scope and 0.3g full-scope bins identified in Table 3.1, and (2) identify enhancements so that the method can be used for the implementation of the IPEEE.

3.2.5.1 Selection of Alternative Success Paths

The EPRI SMM as currently constituted calls for evaluation of a preferred path and an alternative path. The NRC panel that reviewed the EPRI methodology recommended:

"... that a reasonably complete set of potential success paths be set down initially, rather than a very small number, since limiting the number of success paths too quickly can prevent the identification of some plant-level HCLPF insights, and can mask plant differences regarding defense-in-depth. The Panel believes that preliminary analysis to narrow the number of paths to the required two or three should begin with the fuller set, and it recommends that this narrowing be documented in detail."

For IPEEE purposes, it is desirable that, to the maximum extent possible, the alternative path involve operational sequences, systems, piping runs, and components different from those used in the preferred path. The procedure used in the trial application of the EPRI methodology (EPRI NP–6359) provides an acceptable approach for use in selecting success paths (preferred and alternative).

3.2.5.2 Walkdown

Same as Section 3.2.4.1.

3.2.5.3 Relay Evaluation

Same as Section 3.2.4.2.

3.2.5.4 Soil Failures

Same as Section 3.2.4.3.

3.2.5.5 Screening Criteria (Use of Screening Tables)

Same as Section 3.2.4.4.

3.2.5.6 Seismic Input

Same as Section 3.2.4.5.

3.2.5.7 Evaluation of Outliers—HCLPF Calculations

Same as Section 3.2.4.6.

3.2.5.8 Nonseismic Failures and Human Actions

Success paths are chosen based on a screening criterion applied to nonseismic failures and needed human actions. It is important that the failure modes and human actions are clearly identified and have low enough probabilities to not affect the seismic margins evaluation. The screening criteria used in the Maine Yankee margin evaluation (NUREG/CR–4826) addressing both single-train and multi-train systems is an acceptable approach. The redundancies along a given success path should be specifically analyzed and documented when they exist. (In a complementary sense, where a single component is truly "alone" in performing a vital function along a success path, this should be highlighted too). This information will serve to indicate the extent to which a single failure would or would not invalidate the plant's ability to respond safely to a given earthquake level.

3.2.6 Containment Performance

The primary purpose of the evaluation for a seismic event is to identify vulnerabilities that involve early failure of containment functions. These include containment integrity, containment isolation, prevention of bypass functions, and some specific systems depending on a containment design (e.g., igniters, suppression pools, ice baskets). The analyses performed for internal events IPE should be used to determine the scope of systems for the examination.

Each licensee should develop a plan to address containment performance during a seismic event consistent with the above-defined purpose. Additional guidance (no requirements implied) on extending margin-type approaches to obtain containment insights is contained in Budnitz 1991a and 1991b, and Reed, et al., 1990. Some general guidance is provided here based on past PRA experience and some generic capacity estimates of typical components involved in containment systems. From a survey of past PRAs (Amico, 1989), it appears that high-consequence sequences involve gross structural failure of the containment itself or failure of major equipment or structures within the containment at very high accelerations (HCLPF values greater than 0.5g) and isolation failure due to seismically induced relay chatter.

Generally, containment penetrations are seismically rugged; a rigorous fragility analysis is needed only at review levels greater than 0.3g, but a walkdown to evaluate for unusual conditions (e.g., spatial interactions, unique penetration configurations) is recommended. An evaluation of the backup air system of the equipment hatch and personnel lock that employ inflatable seals should be performed at all review levels. Also, some penetrations need cooling, and the possibility and consequences of a cooling loss caused by an earthquake should be considered.

Valves involved in the containment isolation system are expected to be seismically rugged (NUREG/CR-4734). A walkdown to ensure that they are similar to test data and have known high capacities and that there are no spatial interaction issues will suffice. Seismic failures of actuation and control systems are more likely to cause isolation system failures and should be included in the examination. For valves relying on a backup air system, the air system should also be included in the seismic examination.

Components of the containment heat removal/pressure suppression functional system that are not included elsewhere and are not known to have high capacities should be examined. An example of such a component might be a fan cooler unit supported on isolator shims. The walkdown should include examination of such components and their anchorages. Similarly, support systems and other system interaction effects (e.g., relay chatter) should be examined as applicable.

For Mark I and ice condenser containments, it is not feasible to screen out components (e.g., torus, ice basket support) on a generic capacity basis. The potential for accident sequences initiated by a containment functional failure should be examined.

3.3 Optional Methodologies

A licensee may request the staff to review any other systematic examination method to determine its acceptability for IPEEE purposes.

4 ACCEPTABLE METHODOLOGY FOR PERFORMING THE INTERNAL FIRES IPEEE

For purposes of an IPEEE, a Level 1 probabilistic risk assessment (PRA) is considered acceptable to identify potential internal fire vulnerabilities at nuclear power plants. Some fire issues identified in the Fire Risk Scoping Study, (1) seismic/fire interactions, (2) effects of fire suppressants on safety equipment, and (3) control system interactions, should be addressed in the IPEEE. The walkdown procedures of the IPEEE should address the above issues and should be specifically tailored to assess the potential vulnerabilities related to these issues. The licensee should use a plant-specific data base on fire brigade training in the IPEEE to assess the effectiveness of manual fire fighting to determine the response time for the manual fire fighters. The licensee should also show the effectiveness of fire barriers in the IPEEE. The current fire PRA method has its limitations (NUREG/CR–5088, 1989) and significant "engineering judgment" must be brought to bear once the PRA has been accomplished to allow for sensible application of the results. The staff believes that the type of "engineering judgment" needed to interpret the results of a PRA is fully within the competence of most fire-safety experts, including experts within the regulatory staff. Further, despite current limitations in the methodology, a fire "vulnerability search" in the spirit of the Severe Accident Policy Statement and the IPE exercise is feasible, and such a vulnerability search need not wait for the completion of further methodology development. Finally, in meeting the objectives of the IPEEE, it is desirable to focus on relative insights rather than on absolute core damage frequency.

4.1 New Fire PRA Analysis

There are several different approaches for the analysis of fires (NUREG/CR–2300, 1983, NUREG/CR–2815, 1985, NUREG/CR–4840, 1990, and NUREG/CR–5259, 1990). Although not all fire PRAs delineate their analysis steps in exactly the same way, the following steps, in one form or another, should be part of any analysis.

4.1.1 Identify Critical Areas of Vulnerability

The criterion is whether a fire could compromise important safety equipment. Emphasis should be placed on areas where multiple equipment could be compromised, in particular, several trains of redundant equipment to perform the same safety function. Attention should be given to the potential for cross-zone spread of fire and the likelihood that transient fuels might supplement fuels already present in a zone.

4.1.2 Calculate the Frequency of Fire Initiation in Each Area

This calculation is sensitive to location within a larger area, particularly if fuel loading conditions, cross-zone

spreading potential, or other idiosyncrasies are considered. Also, the data base on fires in various areas should be coupled with location-specific information obtained from the plant walkdown and other experience to account for uncertainties.

4.1.3 Analyze for the Disabling of Critical Safety Functions

Determine the likelihood of equipment being disabled by a fire. The areas to be addressed include:

1. Fire growth and spread, including the treatment of hot gases and smoke.

2. Detection/suppression effectiveness and reliability.

3. Component fragility to fire and combustion products.

4. Probability estimates (distributions) for fault tree quantification.

4.1.4 Identify Fire-Induced Initiating Events/ Systems Analysis

Perform the analysis to determine the frequency of fire initiated accident sequences leading to core damage.

4.1.5. Perform Containment Analysis

Perform containment analysis if containment failure modes differ significantly from those found in the IPE internal events evaluation.

Perform in a fashion similar to an internal-initiator PRA.

4.2 Use of an Existing Fire PRA

The use of an existing fire PRA for the internal fires IPEEE is acceptable provided the PRA reflects the current as-built and as-operated status of the plant and the licensee addresses the deficiencies of past PRAs that are identified in the Fire Risk Scoping Study (NUREG/CR–5088). Deficiencies may include the use of low conditional failure probabilities for dampers and penetrations, no consideration of damage from the use of fire suppressants, inappropriate estimates of the effectiveness of manual fire fighting, and no consideration of seismic/fire interactions.

4.3 Optional Methodologies

A licensee may request the staff to review any other systematic examination method to determine its acceptability for IPEEE purposes.

5 ACCEPTABLE METHODOLOGY FOR PERFORMING THE HIGH WINDS, FLOODS, AND TRANSPORTATION AND NEARBY FACILITY ACCIDENT IPEEE

For the purposes of an IPEEE, the staff recommends a progressive screening approach to identify potential vulnerabilities at nuclear power plants due to high winds, floods, and transportation and nearby facility accident. The owners of Trojan and Washington Nuclear Plant 2, who are requested to evaluate the effects of volcanic activities in assessing severe accident vulnerabilities, should determine if the recommended screening approach is applicable to their unique situation.

5.1 Introduction

It is assumed that the IPE for internal events will be in progress or completed when the portion of the IPEEE pertaining to high winds, floods and transportation and nearby facility accident is being performed. Some external events will be addressed in the internal events IPE analyses (e.g., the primary effect of lightning is loss of offsite power, which is included in the internal events analyses); other external events will have been screened from further consideration by the staff. For those external events not in either of these categories, further consideration using the progressive screening approach shown in Figure 5.1 is recommended.

5.2 Analytical Procedure

The steps shown in Figure 5.1 represent a series of analyses in increasing level of detail, effort, and resolution. However, the licensee may choose to bypass one or more of the optional steps as long as the 1975 Standard Review Plan (SRP) (NUREG–75/087) criteria are met or the potential vulnerabilities are either identified or demonstrated to be insignificant.

In general, the containment structure, equipment hatch, personnel air lock, and other penetrations are designed and constructed to have high capacities in resisting the effects of high winds, floods, and overpressure induced by transportation or nearby facility accidents. Therefore, no additional containment performance assessment (beyond that discussed for the seismic portion of the IPEEE in Sections 3.1.1.5, 3.1.2, and 3.2.6) is needed unless a licensee predicts or identifies plant-unique accident sequences different from those determined by the internal events IPE.

5.2.1 Review Plant-Specific Hazard Data and Licensing Bases

All licensees should review the information on plant design hazard and the licensing bases, including the resolution of each event.

5.2.2 Identify Significant Changes Since OL Issuance

All licensees should review the site for any significant changes since the operating license was issued with respect to (1) military and industrial facilities within 5 miles of the site, (2) onsite storage or other activities involving hazardous materials, (3) transportation, or (4) developments that could affect the original design conditions.

5.2.3 Determine if the Plant/Facilities Design Meets 1975 SRP Criteria

All licensees should compare the information obtained from the review discussed in Sections 5.2.1 and 5.2.2 for conformance to 1975 SRP criteria and perform a confirmatory walkdown of the plant. The walkdown would concentrate on outdoor facilities that could be affected by high winds, onsite storage of hazardous materials, and offsite developments. If the comparison indicates that the plant conforms to the 1975 SRP criteria and the walkdown reveals no potential vulnerabilities not included in the original design basis analysis, it is judged that the contribution from that hazard to core damage frequency is less than 10^{-6} per year and the IPEEE screening criterion is met.

Otherwise or if a licensee knows that the 1975 SRP criteria will not be met, it should take one or more of the optional steps given in Sections 5.2.4, 5.2.5, and 5.2.6 to further evaluate the situation.

5.2.4 Determine if the Hazard Frequency is Acceptably Low (Optional Step)

If the original design basis does not meet current regulatory requirements, the licensee may choose to demonstrate that the original design basis is sufficiently low—that is, less than 10^{-5} per year, and the conditional core damage frequency is judged to be less than 10^{-1}.

If the original design basis hazard combined with the conditional core damage frequency is not sufficiently low (i.e., less than the screening criterion of 10^{-6} per year), additional analysis may be needed.

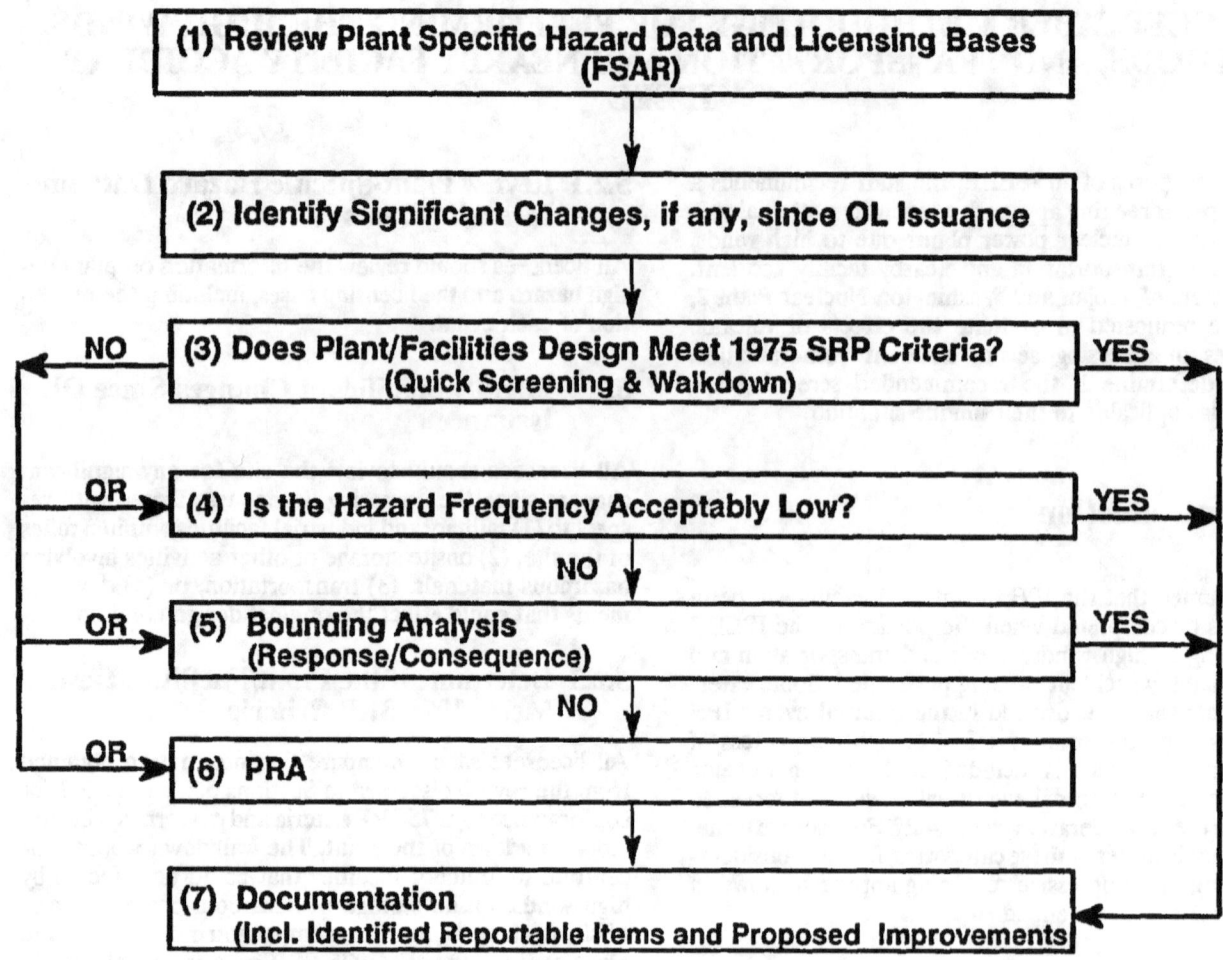

Figure 5.1 Recommended IPEEE Approach for Winds, Floods, and Others

5.2.5 Perform a Bounding Analysis (Optional Step)

This analysis is intended to provide a conservative calculation showing that either the hazard would not result in core damage or the core damage frequency is below the reporting criterion. The level of detail is that level needed to demonstrate the point; judgment is needed for determining the proper level of detail and needed effort.

5.2.6 Perform a Probabilistic Risk Assessment (Optional Step)

A probabilistic risk assessment (PRA) consists of the fol-

lowing key elements: hazard analysis, fragility evaluation, plant systems and accident analysis (event/fault trees), and radioactive material release analysis. The detailed procedure is described in NUREG/CR–2300, NUREG/CR–2815, and NUREG/CR–5259. If the core damage frequency is less than 10^{-8} per year, the event need not be considered further. The level of detail is that level needed to conclude that the core damage frequency is low or to find vulnerabilities.

5.3 Optional Methodologies

A licensee may request the staff to review any other systematic examination method to determine its acceptability for IPEEE purposes.

6 COORDINATION WITH ONGOING PROGRAMS

6.1 Introduction

If unnecessary duplication of effort is to be avoided, coordination with ongoing programs is necessary. The first level of coordination consists of the three major elements related to the implementation of the Severe Accident Policy, that is, coordination of the IPEEE with the internal events IPE and accident management. The second level of coordination consists of the three major elements of the IPEEE, that is, seismic events, internal fires, and high winds, floods, and others. The third level of coordination consists of each major element of the IPEEE, for example, seismic events, and the ongoing programs related to that element.

6.2 Description of Ongoing Programs

6.2.1 IPE Program Related to Internal Events

In Generic Letter 88–20, the NRC requested that the licensee of each plant to perform a systematic examination to identify any plant–specific vulnerabilities to severe accidents and to report the results to the staff. The process was defined as an individual plant examination (IPE). Licensees were requested to proceed with the examinations for internally initiated events only (including internal flooding). Examination of externally initiated events would proceed separately and on a later schedule. However, while performing the IPE for internally initiated events, licensees were advised to document and retain plant-specific data relevant to external events so that they can be readily retrieved in a convenient form when needed for later external event analyses.

6.2.2 Programs Related to External Events

6.2.2.1 Seismic Programs

The following is a brief description of the programs related to seismic events:

1. USI A–17, "System Interactions in Nuclear Power Plants," addresses NRC's concerns regarding the interaction of various systems with regard to whether actions or consequences could adversely affect the redundancy and independence of safety systems. The evaluation of system interactions related to internal events and internal floods is included in the IPE (GL 88–20). The evaluation of spatial system interaction under seismic conditions (the SSE) is included in USI A–46.

2. USI A–40, "Seismic Design Criteria," investigates selected areas of the seismic design process. The staff identified alternative approaches to certain design procedures and modifications to the NRC criteria in the Standard Review Plan to reflect the current state of the art and industry practice. The concern for the seismic capacity of safety-related above-ground tanks (at the SSE) is included in USI A–46.

3. USI A–45, "Shutdown Decay Heat Removal Requirements," has the objective of determining whether the decay heat removal function at operating plants is adequate and if cost-beneficial improvement could be identified. USI A–45 was subsumed in the IPE (GL 88–20); therefore, the external event aspects including the seismic adequacy of the decay heat removal system should be included in the IPEEE.

4. USI A–46, "Verification of Seismic Adequacy of Equipment in Operating Plants" has developed an alternative method and acceptance criteria (to current licensing requirements) to verify the seismic adequacy of equipment in some operating plants with construction permit applications docketed before about 1972. All these plants will be reviewed to the existing safe shutdown earthquake (SSE). The scope of USI A–46 has been expanded to cover the seismic spatial system interaction of USI A–17 and the concern of USI A–40 for the seismic capability of large safety-related above-ground tanks.

5. GI–131, "Potential Seismic Interaction Involving the Movable In-Core Flux Mapping System Used in Westinghouse Plants," was identified because portions of the in-core flux mapping system that have not been seismically analyzed are located directly above the seal table. Failure of this equipment during a seismic event could cause multiple failures at the seal table and could produce an equivalent small-break LOCA.

6. The "Eastern U.S. Seismicity Issue" (formerly the Charleston Earthquake Issue) came about as a result of a U.S. Geological Survey letter in 1982 that pointed out the possibility that large, damaging earthquakes have some likelihood of occurring at locations that had not been considered in licensing decisions. The staff initiated the Seismic Hazard Characterization Project at LLNL, which provided probabilistic seismic hazard estimates for all nuclear power plant sites east of the Rocky Mountains. A similar project was carried out by EPRI for the electric utility industry. The staff's purpose in evaluating the probabilistic studies has been to identify plants in the Central and Eastern United States where past

licensing decisions may have resulted in their being outliers with respect to seismic hazard, that is, the likelihood of exceeding their design bases. As a result of the probabilistic analyses performed, eight plants at five Eastern U.S. sites were identified as being outliers. The IPEEE will provide a resolution for the outlier plants with no need for additional analyses or documentation from the licensees.

6.2.2.2 Internal Fires Programs

The following is a brief description of programs related to internal fires:

1. NUREG/CR-5088, "Fire Risk Scoping Study," identifies some fire issues that had not previously been addressed in the fire PRAs: fire growth code, seismic/fire interaction, fire barrier effectiveness, manual fire fighting effectiveness, effects of fire suppressants on safety equipment, and control system interactions. A plant-specific analysis (including a specifically tailored walkdown) should be performed to assess the actual risk impact of these issues at a plant.

2. GI-57, "Effects of Fire Protection System Actuation on Safety-Related Equipment," assesses the impact of inadvertent actuation of fire protection systems on safety systems. This is one of the issues identified in the Fire Risk Scoping Study. The industry, through EPRI, has a program collecting data on the effects of suppressants on the safety equipment.

3. USI A-45, "Shutdown Decay Heat Removal Requirements," was initiated to determine if the decay heat removal function at operating plants is adequate and if cost-beneficial improvement could be identified. The USI A-45 was subsumed in the IPE (GL 88-20); therefore, the adequacy of the decay heat removal system under internal fires should be addressed in the fire IPEEE.

6.2.2.3 External Flooding Program

GI 103, "Design for Probable Maximum Precipitation (PMP)," is a related external flooding issue. The staff provided the resolution of this issue to all licensees in Generic Letter 89-22, dated Oct. 19, 1989. Specifically, the NRC requested that future plants be designed against a new PMP criterion. For existing plants, the NRC recommended that licensees review the information contained in Generic Letter 89-22 and determine if they need to take additional action. For the IPEEE, the severe accident risk from PMP should be assessed. The licensees should assess the effects of applying this new PMP criterion to their plants in terms of onsite flooding and roof ponding to determine whether that would lead to severe accidents.

6.3 Approach on Coordination with Ongoing Programs

If duplication of effort by the staff and licensees is to be avoided, it is important that the above ongoing programs be coordinated.

6.3.1 Coordination With Internal Events Program (IPE)

The coordination between the internal events IPE and the IPEEE can be categorized into three phases: preanalyses planning, plant modifications, and accident management.

6.3.1.1 Preanalyses Planning

Considerations include (1) definition of elements and their boundaries, (2) walkdown procedures and spatial interactions, and (3) composition of the peer review group. It is likely that the IPE will precede the IPEEE. Careful planning, taking into account the above considerations, will avoid a duplication of effort by the licensee.

6.3.1.2 Plant Modifications

Since the IPE and the IPEEE are likely to be performed separately, it is imperative to examine the impact of modifications identified during the IPE on external events and vice versa. The staff examined several PRAs that included both internal events and external events (Bohn, 1989) to identify possible interactions. Highlights of the major findings in the seismic area (which to some extent are applicable to fire), are the following:

1. In general, the modifications proposed as a result of the internal events analysis would not adversely affect the seismic or fire risk, provided the modifications do not become weak links.

2. In general, the modifications made could potentially contribute to an increase in risk at the plant in the following ways:

   a. Many of the modifications proposed may involve adding valves or suction lines to existing systems. Thus, the possibility of violating the pressure boundary and creating a potential diversion exists if the modification were to fail during an earthquake. Also, modifications may involve routing different trains of electrical power or power from adjacent units. The possibility exists that the circuitry could be designed in such a way that failure of non-safety-related electrical components could actually defeat the circuitry that was desired to provide redundancy, and

b. The possibility exists that inadequate anchorage could defeat the planned redundancy during a seismic event.

3. The potential adverse effects of the modifications include:

    a. Poor accessibility for maintenance,

    b. Stiffening of systems leading to higher stress due to thermal cycles during normal plant operation.

The cited study (Bohn, 1989) provides specific examples of modifications and their effects on other initiating events.

6.3.1.3 Accident Management

Guidance on the integration of findings from the IPEEE and accident management is being developed (SECY-89–308, Oct. 1989).

6.3.2 Coordination Among External Events Programs

The issue of integration between external events primarily involves interactions between seismic events and fires and seismic events and floods. Seismically induced fires and floods are to be addressed as part of the IPEEE. The effects of seismically induced fires and the impact of inadvertent actuation of fire protection systems on safety systems should be addressed. The effects of seismically induced external flooding and internal flooding on plant safety should be included. The scope of the evaluation of seismically induced floods, in addition to that of the external sources of water (e.g., tanks, upstream dams), should include the evaluation of some internal flooding consistent with the discussion in Appendix I of EPRI NP-6041. The coordination between the seismic and the fire or flood analysts should be based on the following:

1. The seismic analysts should generally search for and identify the initiating events (certain specific seismically initiated failures of equipment or structures) that can cause fires or floods, and

2. The seismic and fire or flood analysts should also discuss other concurrent seismically induced failures or possible effects on human actions and then, proceed to complete the rest of the IPEEE analysis.

The coordination should include a meeting, prior to seismic walkdown, in which the fire and flood analysts discuss the key issues, how the analysis will be done, and what to look for. The fire or flood analyst may need to participate in parts of the seismic walkdown or revisit the areas iden-

tified during the seismic walkdown to grasp the issues from the seismic-capacity point of view.

6.3.3 Coordination With Seismic Programs

A number of programs related to seismic events requiring licensee action have been identified. Many of these programs have arisen as a result of the changing perception of hazards and revisions in the design and qualification criteria. There are two categories of seismic programs as they relate to the seismic IPEEE. The first category involves programs, e.g., USI A-45, "Shutdown Decay Heat Removal Requirements", GI-131, "Potential Seismic Interaction Involving the Movable In-Core Flux Mapping System Used in Westinghouse Plants," and "the Eastern U.S. Seismicity Issue," that have been subsumed into the IPE/IPEEE. USI A-45 and GI-131 should be specifically addressed as part of the seismic IPEEE. The Eastern U.S. Seismicity Issue requires no additional licensee actions or reporting. The second category involves programs (e.g., USI A-46, "Verification of Seismic Adequacy of Equipment in Operating Plants,") that can be coordinated with the seismic IPEEE. The coordination of these programs with the seismic IPEEE is most beneficial in reducing the resources expended by the licensee and the staff.

6.3.3.1 USI A-45 and GI-131

The methodology used in the seismic IPEEE can also be used to address USI A-45 and GI-131. The systems and components for addressing USI A-45 will have been determined by the internal events IPE, and the purpose of the seismic IPEEE is to identify any significant and unique seismic vulnerabilities in the decay heat removal function. In addition, the seismic IPEEE will evaluate the potential seismic interaction of the movable in-core flux mapping system used in Westinghouse plants.

Capacities of decay heat removal components can be established using either the fragility analysis (FA) or conservative deterministic failure margin (CDFM) approaches depending upon the methodology chosen to implement the seismic IPEEE. Thus resolution of these issues can be easily accomplished during the seismic IPEEE evaluation.

The potential interaction between the seal table and non-Category I seismic systems associated with the movable in-core flux mapping system can be identified during the seismic walkdown of the IPEEE. If needed, the component capacities or consequences of component failure can be evaluated using the same procedures that are used in the seismic IPEEE.

6.3.3.2 The Eastern U.S. Seismicity Issue (The Charleston Earthquake Issue)

As a result of work carried out to resolve the Eastern U.S. Seismicity Issue (Charleston Earthquake Issue), probabilistic seismic hazard estimates exist for all nuclear power plants east of the Rocky Mountains. These should be used directly by any licensee in that region opting to satisfy the seismic IPEEE with a seismic PRA. The hazard estimates also played a key role in determining the review level earthquake used in the seismic margins methodology option. The IPEEE will provide a resolution for this issue without requiring additional analyses or documentation from licensees.

6.3.3.3 USI A-46

Implementation of the USI A-46 program involves plants with construction permit applications docketed before about 1972. The USI A-46 plants thus form a subset of all the nuclear power plants in the U.S. that are requested to perform the seismic IPEEE.

The most efficient way to address the ongoing seismic programs for USI A-46 plants is to conduct the A-46 review and walkdown to gather relevant information for the seismic IPEEE. In order to facilitate this approach, the activities of USI A-46 and the seismic IPEEE need to be coordinated, and the plant walkdown needs to be well planned. Several inherent differences between the A-46 program and the seismic IPEEE should be noted at the outset before attempting to coordinate the two programs.

First, the objectives are quite different. The USI A-46 program has licensing implications on plant operation; this program will assess and ensure the seismic ruggedness of safety-related equipment in a plant to withstand the SSE. The seismic IPEEE, on the other hand, generally tries to identify plant vulnerabilities when subjected to earthquake levels higher than the SSE design basis.

Second, the scope of the reviews are different. USI A-46 is concerned with only one success path (with some requirement on equipment redundancy) of equipment needed to bring the plant to safe shutdown in the event of an earthquake and maintain it there for at least 72 hours. The scenario considers an earthquake of the SSE level with a possible loss of offsite power because of this earthquake. The probabilities of a seismically induced LOCA (small or large) and a high-energy line break (HELB) occurring are judged to be low enough that their consideration at this earthquake level is not warranted. Piping, tubing, and structures will be examined during a walkdown only if they have the potential to cause seismic interaction with the equipment reviewed and cause damage to this equipment. The review of above-ground tanks (as part of USI A-40) is an exception. The seismic IPEEE is concerned with the vulnerabilities of the whole plant, not just the equipment. Also, evaluations are generally made at levels above the design basis. At this level of earthquake, seismically induced LOCAs are considered, and mitigating systems and equipment to address this initiator are reviewed. Therefore, even if the EPRI seis-

mic margins methodology is utilized to implement the seismic IPEEE (since it is quite similar to the USI A-46 evaluation), it would need additional equipment to be reviewed over that required for implementing USI A-46.

Third, the levels of review and walkdown are different. The Seismic Qualification Utility Group (SQUG) and EPRI have developed a detailed Generic Implementation Procedure (GIP) for the USI A-46 review and walkdown that was reviewed by the NRC staff, and a Safety Evaluation Report (SER) was issued. The GIP should be followed in performing the USI A-46 review and walkdown. The guidelines associated with the seismic PRA or seismic margins methodology are not as specific as those in the GIP. To illustrate this point, in the walkdown review of expansion anchor bolts, GIP calls for the use of a wrench test for the bolt tightness check, whereas the margins walkdown ensures only that the anchor bolts are adequate to hold down the equipment as designed with no specific testing requirements to confirm anchor capacity. The completion of the seismic IPEEE does not automatically mean that the USI A-46 review is satisfactorily completed.

There may be overlaps or differences in the equipment scope for USI A-46 and the seismic IPEEE. For equipment that is within the scope of USI A-46 or the seismic IPEEE only, it is clear that either GIP or IPEEE guidelines, respectively, should apply. For the overlapping equipment, the efficient approach is to use the GIP for both walkdowns; however, the IPEEE should use the review level earthquake. Caveats and interaction (such as flooding) provisions of EPRI NP-6041 should be observed.

In summary, it is recommended that licensees coordinate the information collection for the USI A-46 and seismic IPEEE review and walkdown in order to minimize or avoid duplication of effort by the licensees and staff. Care should be exercised in the coordination to ensure that the objectives of both programs are fulfilled. Coordination of the two programs has been shown to be feasible in the trial evaluation of the Hatch plant using the EPRI seismic margins methodology.

6.3.4 Coordination With Other Issues

In addition to the specific USIs and GIs discussed above, if, during its IPEEE, a licensee (1) discovers a notable vulnerability that is topically associated with any other USI or GI and proposes measures to dispose of the specific safety issue or (2) concludes that no vulnerability exists at its plant that is topically associated with any USI or GI, the staff will consider the USI or GI resolved for a plant upon review and acceptance of the results of the IPEEE. The following should be discussed:

a. The ability of the methodology to identify vulnerabilities associated with the USI or GI being addressed.

b. The contribution of each USI or GI to core damage frequency or unusually poor containment performance, including sources of uncertainty when PRA is used.

c. The technical basis for resolving the issue.

7 PEER REVIEW

In Supplement 4 to Generic Letter 88–20, the staff requests that each licensee conduct a peer review by individuals who are not associated with the initial evaluation to ensure the accuracy of the documentation and to validate both the IPEEE process and its results. The submittal should contain, as a minimum, a description of the review performed, the results of the review team's evaluation, and a list of the review team members.

The purpose of the peer review is twofold. The first purpose is to provide quality control and quality assurance to the IPEEE process. Independence of the review team is desirable because it reflects a quality control and quality assurance attitude. This does not imply that the staff is seeking a review or document control similar to that specified in Appendix B to 10 CFR Part 50. The staff does seek to ensure that the IPEEE process produces reliable, factual information. If it is necessary to use a reviewer who is not totally removed from the plant-specific IPEEE process, the licensee should be confident that the reviewer can be objective and capable of providing a critical review. The "in-house" team *can be supplemented* by outside consultants as determined appropriate by the licensee.

The second purpose of the review relates to the importance of having utility personnel cognizant of the IPEEE. The maximum benefit to the utility would occur if the combination of persons involved in the original analysis and peer review, taken as a group, provide both a cadre of utility personnel to facilitate the continued use of the results and the expertise in the methods to ensure that the techniques have been correctly applied. The staff expects all utilities to have in-house personnel who have the most expert knowledge of their plants, system configurations, and operating practices and procedures.

The staff recommends that the peer review team have combined experience in the areas of systems engineering and specific external events. For example, the seismic peer review team should have combined experience in the areas of sytems engineering, seismic capacity engineering, and seismic PRAs or seismic margins methodologies.

7.1 Seismic Related Insights

Three trial plant applications of the seismic margins methodologies have demonstrated that considerable judgment is involved in applying these methods, and that peer review groups have considerably aided this judgment and have also enhanced the overall credibility of the studies (Davis, 1990).

A meeting of the peer review team with the seismic review team prior to the plant walkdown will provide insights into the appropriateness of the proposed plant walkdowns. Also, the peer review team's endorsement of the final results will add credibility to the HCLPF estimates.

8 DOCUMENTATION AND REPORTING

The IPEEE should be documented in a traceable manner to provide the basis for the findings. This can be dealt with most efficiently by a two-tier approach. The first tier consists of the results of the examination, which will be reported to the NRC for review. The second tier is the documentation of the examination itself, which should be retained by the licensee for the duration of the license unless superseded.

The information submitted to the NRC should be organized and presented in accordance with Appendix C. The submittal may enable many issues to be dealt with in the IPEEE review. Pertinent issues are discussed in Section 6. For some issues, for example, USI A-46, a detailed documentation requirement exists, and it should be followed in the broad framework of IPEEE submittals. Specific information relevant to particular issues, e.g., USIs and GIs, should be identified.

8.1 Information Submitted to the NRC

A detailed list of information to be submitted to the NRC is provided in Appendix C.

8.2 Information Retained for Audit

Retained documentation should include applicable event trees and fault trees, current versions of the system notebooks (if applicable), walkdown reports, and the results of the examination. In general, all documents essential for a practitioner in the field to understand what was done in the IPEEE should be retained. In addition, the manner in which the validity of these documents has been ensured should be documented. If credit is allowed in the IPEEE for any actions taken by the operators, the licensee should have established plant procedures to be used by the plant staff responsible for managing a severe accident should one occur. Procedures should provide assurance that the operators can and will take the proper action.

9 REFERENCES

Amico, P., "Containment Considerations for the Use of the Seismic Margins Methods for Risk Screening," Letter Report, Science Applications International Corporation, June 30, 1989.

Beckjord, E. (NRC), memorandum to L. Shao (NRC), Subject: External Events Steering Group, December 21, 1987.

——, memorandum to L. Shao (NRC), Subject: External Events Steering Group, May 31, 1988.

Bohn, M., "Status Report on Issues Related to Internal/External Event Interaction and Decay Heat Removal Requirements for IPEs," Draft, Sandia National Laboratory, May 1989.

Budnitz, R., et al., "Extending a HCLPF-Based Seismic Margin Review to Analyze the Potential for Large Radiological Releases and the Importance of Human Factors and Non-Seismic Failures," Draft, Future Resources Associates, Inc., March 1987.

Budnitz, R., letter to C. McCracken (NRC), Subject: Modification to Memorandum of 12 July 1988, January 3, 1989.

——, "Enhancing the NRC and EPRI Seismic Margin Review Methodologies to Analyze the importance of Non-Seismic Failures, Human Errors, Opportunities for Recovery, and Large Radiological Releases, Draft 2, September 1990.

Chery, D., (NRC), memorandum to D. Moeller and D. Okrent (ACRS), Subject: Hydrologic Engineering Presentation to Combined Meeting of the ACRS Subcommittees on Site Evaluation and Extreme Events Phenomena, October 9, 1985.

Code of Federal Regulations, Title 10 "Energy" (10 CFR), U.S. Government Printing Office, Washington, D.C., revised periodically.

Daily, C. (NRC), memorandum to R. Savio and M. Stella (ACRS), Subject: Assessment of Issue Concerning Operating Reactors: Lightning Induced Reactor Events, August 2, 1989.

Davis, P., "A Peer Review of Two Seismic Margin Assessments as Applied to the Hatch Nuclear Power Plant", Proceedings of Third Symposium on Current Issues Related to Nuclear Power Plant Structures, Equipment and Piping, Orlando, Florida, December 1990.

Electric Power Research Institute, EPRI NP-6041, "A Methodology for Assessment of Nuclear Power Plant Seismic Margin," October 1988.

——, EPRI NP-6359, "Seismic Margin Assessment of the Catawba Nuclear Station," Vols. 1 and 2, April 1989.

——, EPRI NP-6395-D, "Probabilistic Seismic Hazard Evaluation at Nuclear Plant Sites in the Central and Eastern United States: Resolution of the Charleston Issue," April 1989.

Moore, D., et al., "Results of the Seismic Margin Assessment of Hatch Nuclear Power Plant," Proceedings of Third Symposium on Current Issues Related to Nuclear Power Plant Structures, Equipment and Piping, Orlando, Florida, December 1990.

Orvis, D., et al, "Seismic Margin Review of Plant Hatch Unit 1: System Analysis," LLNL Report No. ULRL-CR-104834, August 1990.

Rasin, W. (NUMARC), letter to W. Minners (NRC), Subject: Final Industry Comments on Draft Generic Letter 88-20, Supplement 4, "Individual Plant Examination of External Events (IPEEE) for Severe Accident Vulnerabilities," and Draft NUREG-1407, "Procedural and Submittal Guidance for the IPEEE," October 10, 1990.

Reed, J., et al. "Recommended Seismic IPE Resolution Procedure," Proceedings of Third Symposium on Current Issues Related to Nuclear Power Plant Structures, Equipment and Piping, Orlando, Florida, December 1990.

Shao, L., et al., "Consideration of External Events in Severe Accidents," Proceedings of Third Symposium on Current Issues Related to Nuclear Power Plant Structures, Equipment and Piping, Orlando, Florida, December 1990.

USNRC, AEOD Engineering Evaluation Report, "Lightning Events at Nuclear Power Plants," AEOD-E605, April 1986.

——, Generic Letter 88-20, "Individual Plant Examination for Severe Accident Vulnerabilities—10 CFR 50.54(f)," November 23, 1988.

——, Generic Letter 88-20, Supplement No. 1. "Initiation of the Individual Plant Examination for Severe Accident Vulnerabilities—10 CFR 50.54(f)," August 29, 1989.

——, Generic Letter 88-20, Supplement No. 4, "Individual Plant Examination of External Events (IPEEE) for Severe Accident Vulnerabilities—10 CFR 50.54(f)," draft for comment, July 23, 1990.

——, Generic Letter 88-20, Supplement No. 4, "Individual Plant Examination of External Events (IPEEE) for Severe Accident Vulnerabilities—10 CFR 50.54(f)," final, April, 1991.

——, Generic Letter 89–22, "Resolution of Generic Safety Issue No. 103, 'Design for Probable Maximum Precipitation'," October 19, 1989.

——, NUREG–1032, "Evaluation of Station Blackout Accidents at Nuclear Power Plants," June 1988

——, NUREG–1150, "Severe Accident Risks: An Assessment for Five U.S. Nuclear Power Plants," Vols. 1 and 2, June 1989.

——, NUREG–1335, "Individual Plant Examination: Submittal Guidance," final report, August 1989.

——, NUREG–75/087, "Standard Review Plan for the Review of Safety Analysis Report for Nuclear Power Plants," LWR edition, December 1975.

——, NUREG/CR–0098, "Development of Criteria for Seismic Review of Selected Nuclear Power Plants," May 1978.

——, NUREG/CR–2300, "PRA Procedures Guide," January 1983.

——, NUREG/CR–2815, "Probabilistic Safety Analysis Procedures Guide," Vols. 1 and 2, August 1985.

——, NUREG/CR–4334, "An Approach to the Quantification of Seismic Margins in Nuclear Power Plants," August 1985.

——, NUREG/CR–4482, "Recommendations to the Nuclear Regulatory Commission on Trial Guidelines for Seismic Margin Reviews of Nuclear Power Plants," March 1986.

——, NUREG/CR–4659, "Seismic Fragility of Nuclear Power Plant Components, Phase I," Vol. 1, June 1986.

——, NUREG/CR–4659, "Seismic Fragility of Nuclear Power Plant Components, Phase II, Motor Control Center, Switchboard, Panelboard and Power Supply," Vol. 2, December 1987.

——, NUREG/CR–4659, "Seismic Fragility of Nuclear Power Plant Components, Phase II, Switchgear, I&C Panels (NSSS) and Relays," Vol. 3, February 1990.

——, NUREG/CR–4734, "Seismic Testing of Typical Containment Piping Penetration Systems," December 1986.

——, NUREG/CR–4826, "Seismic Margin Review of the Maine Yankee Atomic Power Station," Vols. 1–3, March 1987.

——, NUREG/CR–4840, "Recommended Procedures for the Simplified External Event Risk Analysis for NUREG–1150," Sandia National Laboratory, September 1989.

——, NUREG/CR–5042, "Evaluation of External Hazards to Nuclear Power Plants in the United States," December 1987.

——, NUREG/CR–5042, Supplement 1, "Evaluation of External Hazards to Nuclear Power Plants in the United States—Seismic Hazard," April 1988.

——, NUREG/CR–5042, Supplement 2, "Evaluation of External Hazards to Nuclear Power Plants in the United States—Other External Events," February 1989.

——, NUREG/CR–5076, "An Approach to the Quantification of Seismic Margins in Nuclear Power Plants: The Importance of BWR Plant Systems and Functions to Seismic Margins," May 1988.

——, NUREG/CR–5088, "Fire Risk Scoping Study," January 1989.

——, NUREG/CR–5250, "Seismic Hazard Characterization of 69 Nuclear Power Plant Sites East of the Rocky Mountains," Vols. 1–8, January 1989.

——, NUREG/CR–5259, "Individual Plant Examination for External Events: Guidance and Procedures," draft, March 1989.

——, NUREG/CR–5270, "Assessment of Seismic Margin Calculation Methods," March 1989.

——, NUREG/CR–5477, "An Evaluation of the Reliability and Usefulness of External Initiator PRA Methodologies," January 1990.

——, NUREG/CR–5499, "Guidance on Relay Chatter Effects," 1991.

——, NUREG/CR–5501, "Selection of Review Level Earthquake for Seismic Margin Studies Using Seismic PRA Results," June 1991.

——, "Policy Statement on Severe Reactor Accidents," *Federal Register*, Vol. 50, p. 32138, August 8, 1985.

——, SECY 88–147, "Integration Plans for Closure of Severe Accident Issues," May 25, 1988.

APPENDIX A
REVIEW LEVEL EARTHQUAKE

APPENDIX A
REVIEW LEVEL EARTHQUAKE

The seismic margins methodology was designed to demonstrate sufficient margin over the Safe Shutdown Earthquake (SSE) to ensure plant safety and to find any "weak links" that might limit the plant shutdown capability to safely withstand a seismic event larger than the SSE or lead to seismically induced core damage. The methodology involves the screening of components based on their importance to safety and seismic capacity. The seismic margins method utilizes two review or screening levels geared to peak ground accelerations of 0.3g and 0.5g. In areas of low to moderate seismic hazard, most plants that have been evaluated using PRAs or margins studies have been shown to have HCLPFs at or below 0.3g. Past experience indicates that, at the 0.3g screening level, a small number of "weak links" are likely to be identified, efficiently defining the dominant contributors to seismically induced core damage. It is the staff's judgment that the use of a 0.3g review level earthquake for most of the nuclear power plant sites in the Central and Eastern United States (east of the Rocky Mountains) would serve to meet the objectives of the IPEEE.

All sites east of the Rocky Mountains, however, are not subject to the same level of earthquake hazard. The recent studies by LLNL (NUREG/CR–5250) and EPRI (EPRI NP–6395–D) show significant differences depending on location and specific site conditions. Because the two studies do not necessarily agree with each other, it was deemed necessary to use both studies in determining which review level earthquake should be assigned to each site. Hazard comparisons were made using the median, 85th percentile, and mean from the site-specific results provided by the LLNL and EPRI studies. Based on the sensitivity tests and engineering and seismological judgment, the staff has defined the review level earthquake for each site (0.3g, 0.5g, or reduced scope) in Table 3.1. A second criterion, plant design basis, was used to subdivide the 0.3g bin. The subdivision, based on a composite conditional probability of exceeding the SSE for each nuclear power plant, resulted in plants within the 0.3g bin being assigned a full-scope or focused-scope review.

The sites in the Western United States (west of the Rocky Mountains) are treated differently. Those sites in coastal California where the seismic hazard is much higher and the resulting design bases are greater than 0.5g cannot make use of the margins methodology. The other plant sites in the West should use a 0.5g review level earthquake unless it can be demonstrated that the seismic hazard level at a particular plant site is consistent with the seismic hazard at the 0.3g bin plant sites east of the Rocky Mountains. Western sites that show such a consistency in seismic hazard will conduct the full-scope 0.3g margins

review. The results of the binning for the plants in the Western United States are presented in Table 3.2.

The rationale for the selection of the review level earthquakes (RLEs) and the grouping of the plants east of the Rocky Mountains is discussed below.

A.1 Introduction

The specification of a review level earthquake (RLE) for use in carrying out an individual plant examination for external events (IPEEE) has been a complex problem involving the search for consistency. It would be preferable if the selection of the RLEs were completely consistent with the individual plant examination (IPE) for internal events and the inherent strengths of the seismic margins methodologies, but it is very difficult to satisfy all of these elements in any rigorous quantitative sense. Thus, for example, attempting to equate the review level earthquake to the reporting criteria in the IPE (mean sequence frequency leading to core damage of 10^{-6} per year) is fraught with difficulties because of the large uncertainties in numerical estimates of seismically induced core damage, the inappropriateness of a comparison between numerical estimates of seismically and internally induced core damage (the source and treatment of uncertainty can be quite different), and the inherent difficulties in relating the output of a seismic margins study (HCLPF) to estimates of core damage frequency. For some of the same reasons, it was recognized that external initiators, including earthquakes, need not necessarily be treated in the same manner as internal initiators in implementing the Severe Accident Policy. It should be noted that the RLE defines a reporting level. A HCLPF value lower than RLE does not necessarily represent a plant vulnerability. However, the licensee should assess the significance of HCLPF values lower than RLE and take any necessary actions and make other improvements that are deemed appropriate by the licensee.

A.2 General Evaluation Procedure

A.2.1 Data Evaluated

The staff has recommended three review level earthquakes to be used when applying the seismic margins methodology to nuclear power plants east of the Rocky Mountains for the IPEEE. The review levels or "bins" are 0.5g, 0.3g, and a reduced-scope level. The basic information used was the Lawrence Livermore National Laboratory (LLNL) hazard study (NUREG/CR–5250) and the Electric Power Research Institute (EPRI) hazard study (EPRI NP–6395–D). These studies represent state-of-the-art estimates of seismic hazard. Because the two studies do not necessarily agree with each other, it was deemed necessary to use them both in determining which bin a particular site belonged in.

In the LLNL study (NUREG/CR-5250), it was noted that, for some sites, the mean estimates of seismic hazard were dominated by the input of one ground motion expert (No. 5). This dominance was caused by the low attenuation, high uncertainty, and relatively high motion on rock found in this expert's input. This input has received a great deal of attention, and some have argued that it is inconsistent with the data. The staff requested LLNL (as a sensitivity study) to calculate the hazard at nuclear power plant sites east of the Rocky Mountains leaving out the input of this expert.

Data from the Saguenay Event in Quebec, Canada (November 1988), the largest earthquake in eastern North America in 50 years, appears to be quite different from previous data sets and has not helped to resolve the controversy. At this time, in order to avoid relying exclusively on the LLNL results that include the input of expert No. 5, the staff is treating the LLNL hazard estimates based on the other four ground motion experts as a separate study when binning nuclear power plant sites for IPEEE.

A.2.2 Comparison Procedure

Hazard comparisons were made using the mean, median, and 85th percentile from the site-specific results provided by the LLNL and EPRI studies. Each of these pieces of information represents a different way of characterizing the distribution of seismic hazard estimates at each site as determined by a particular study.

Mean: The mean is a commonly used statistic that can be assigned actuarial significance. However, because of the skewed nature of the distribution, it is also a highly unstable (with respect to methodology and input assumptions) view of hazard. The mean is highly sensitive to the characterization of the extremes of the distribution.

Median: The median is more stable than the mean and shows the greatest agreement between the LLNL and EPRI studies. However, it is only the 50th percentile of the hazard and is insensitive to the extent of uncertainty.

85th Percentile: An alternative candidate to the mean is the 85th percentile. It reflects uncertainty in that it indicates the breadth of the distribution, but it is less sensitive to extreme outliers.

A.2.3 Weighting Criteria

In the past, great emphasis has been placed on the likelihood of exceeding peak ground acceleration (PGA). In this evaluation, site hazard comparisons were made using

response spectra and PGA. The likelihoods of exceeding spectral response accelerations in the 2.5 to 10 Hz range were examined because these frequencies are more closely related to the types of motion that could cause damage at nuclear power plants. Unit weights (2/7th each) were assigned to the likelihoods of exceeding spectral response ordinates at 2.5, 5, and 10 Hz. One-half unit weight (1/7th) was assigned to the likelihood of exceeding the PGA.

A.2.4 Ranking Criteria

Emphasis was placed on the relative ranking of sites with respect to other sites using the same seismic hazard study, statistic, and ground motion measures. Extensive use was made of a clustering methodology developed by LLNL for the NRC (Bernreuter et al., 1989a, 1989b). For a given hazard study, statistic, ground motion measure and reference level, this methodology divides the ensemble of sites into groups so that the sites in any one group are "close" to each other with respect to seismic hazard. For example, the sites may be divided into groups based on mean estimates of exceeding 0.5g PGA from the EPRI study or median estimates of exceeding the 2.5 Hz spectral response (associated with the NUREG/CR-0098 response spectrum anchored at 0.3g) from the LLNL five-expert study. Although there were a fixed number of groups, no minimum number of sites were in a group, and indeed some groups contained only one site.

A.2.5 Spectral Shape

The spectral shape associated with the 0.3g screening level was assumed to be the median NUREG/CR-0098 spectrum anchored at 0.3g. There has been some discussion that the screening level should actually be associated with a somewhat higher ground motion (the Seismic Qualification Utility Group (SQUG) bounding spectrum) but in this relative comparison, the use of this alternative spectrum would make little or no difference.

A.3 Specific Binning Procedure

A.3.1 Initial Binning Evaluation

As the first step, sites that consistently fell into the group that had the highest likelihood of exceeding the 0.3g NUREG/CR-0098 5% damped median spectrum were conditionally assigned to the 0.5g bin. Sites that fell into the group that had the lowest likelihood of exceeding the 0.3g NUREG/CR-0098 5% damped spectrum were assigned to the reduced-scope bin.

The ground motion measure compared was the weighted combination of 2.5 Hz, 5 Hz, 10 Hz and PGA. The individual consistency criteria used were:

1. Agreement among the LLNL five-expert, LLNL four-expert, *and* EPRI studies, and

2. Agreement between the median *and* either mean *or* 85th percentile statistics.

This resulted in a comparison of nine separate hazard groupings (three pieces of information for each of the three studies).

For example, if a particular site fell in the top group (0.5g bin) for all of the criteria except the EPRI median, it remained in the 0.3g bin. The conclusions must be supported by all the hazard studies. On the other hand, if a particular site fell in the bottom group for all of the criteria except for the LLNL four- and LLNL five-expert mean estimates, it was included in the reduced-scope bin. Only one measure of uncertainty, mean or 85th percentile, needs to be satisfied.

A.3.2 0.3g Bin Subdivision

The staff investigated the potential for using the seismic design basis as a parameter for making the initial binning assignments. There was insufficient technical basis for its use; thus it was not used for the initial binning. However, when combined with hazard and engineering judgment, the use of the seismic design provided a basis for an overall cost-effective reduction in the scope of the 0.3g margins review. The staff repeated the process that was used to obtain the initial binning with the sole change that instead of factoring in only the seismic hazard, the seismic hazard and the seismic design basis were used.

Composite conditional probabilities were obtained for the three seismic hazard curves (EPRI, LLNL with four experts, and LLNL with five experts) and the three statistical measures of the hazard curves (mean, median, and 85%)—nine separate probabilities for each site. A composite conditional probability was formed by adding the weighted conditional probabilities of exceeding the uniform hazard spectra at a particular ground motion frequency; i.e., the intersection of the plant-specific seismic design spectrum for the particular frequencies with the uniform hazard spectra yields the conditional probability. The frequencies were those used for the initial binning (2.5Hz, 5Hz, 10Hz, and PGA). The were also weighted the same (2/7, 2/7, 2/7, and 1/7).

Using the same agreement criteria as in the initial binning, six sites were identified, i.e., consistently fell into the top group. These are listed in Table 3.1 to do the full-scope 0.3g seismic margins review.

As a "sanity check" of this approach, the list derived from this approach was compared to the list derived from seismic hazard alone. The six full-scope plants were among the top ten seismic hazard sites.

The staff's resolution of the Eastern U.S. Seismicity Issue (Charleston Earthquake Issue) has identified eight plants at five sites as outliers. The staff determined that these plants should be assigned to the full-scope category. This action added a single additional plant, Arkansas Nuclear One, Unit 1, to the list derived on the basis of seismic hazard and seismic design.

The candidates for the 0.5g and reduced-scope bins were then subjected to additional evaluation by the staff.

A.3.3 Subsequent Binning Evaluations

The candidates for the 0.5g bin were first examined to provide some assurance that, although the hazard was relatively high, it was high enough to warrant inclusion in this bin.

As a test, it was considered appropriate that a site belonged in the 0.5g bin if a hypothetical nuclear power plant at that site was assumed to have a HCLPF of 0.3g and the mean annual core damage frequency associated with that hypothetical plant was 10^{-5} or higher. The work cited in NUREG/CR–5501 showed that the mean annual core damage frequency was roughly an order of magnitude lower than the mean annual likelihood of exceeding the plant HCLPF and very roughly equal to the median annual likelihood of exceeding the plant HCLPF.

Based on these estimates, the staff assumed that inclusion in the 0.5g bin would be supported if:

1. The mean or 85th percentile annual likelihood of exceeding the 0.3g spectrum from all three studies was 10^{-4} or greater, and

2. The median annual likelihood of exceeding the 0.3g spectrum from all three studies was 10^{-5} or greater.

This evaluation should be viewed as a "sanity check"; it should not be viewed as a plant-specific statement on core damage frequencies. The reasons are:

1. The uncertainty and generic nature associated with the correlation in NUREG/CR–5501,

2. The use of spectral estimates rather than peak ground acceleration,

3. The inclusion of the 85th percentile estimates, and

4. All the previously mentioned problems associated with bottom line numbers.

Finally, the staff examined the candidates for the 0.5g and reduced-scope bins to assure itself that the classification made good seismological sense and there was no need to include additional sites in these bins. In conjunction with this examination, limited sensitivity tests were also carried out to determine the impact of slight relaxations in the consistency criteria.

A.4 Binning of Sites — Results

A.4.1 Reduced-Scope Margins Methodology Bin

The consistency criteria outlined in Section A.3.1 were slightly modified to identify sites for the reduced-scope bin. The two bottom median groups were included rather than the bottom group alone. When this was done, five sites (South Texas, Comanche Peak, Waterford, River Bend, and Crystal River) were identified as belonging to the reduced-scope bin.

Also added to this bin were several sites for which no EPRI calculations were available but were in the bottom groups in both the LLNL four- and five-expert studies. They are Duane Arnold, Big Rock Point, Grand Gulf, St. Lucie, and Turkey Point. The ten candidate sites in the reduced-scope bin lie in areas of low seismic hazard along or near the Gulf and Florida coasts and in the upper Midwest.

A.4.2 0.5g Bin

As a result of the evaluations cited above, two sites (Pilgrim and Seabrook) were identified as belonging in the 0.5g bin.

A.4.3 0.3g Bin

All sites not identified as belonging in the 0.5g or reduced-scope bins were assigned to the 0.3g bin.

A.4.4 Other Considerations

The grouping was made assuming that each location was associated with one site condition (rock or varying depths of soil). Some twelve plant sites east of the Rocky Mountains whose main Category I structures are located on rock also have some Category I structures or components located on shallow or intermediate depths of soil. Since shallow soil, less than about 80 feet thick, can significantly amplify ground motion, these sites should perform soil amplification studies to determine the effect.

In particular, for four of the sites included in the 0.3g bin (on the basis of their primary site conditions), the hazard for structures or components on the secondary site conditions is equal to or higher than the hazard associated with those plants in the 0.5g bin. Licensees should, if the site-specific analysis indicates, use the 0.5g screening tables for elements affected by soil amplification. Similarly, for one site in the reduced-scope bin, site-specific analysis should be carried out to determine the effects on those elements affected by soil amplification.

APPENDIX B

COMPARISON BETWEEN A REDUCED-SCOPE
AND FULL-SCOPE SEISMIC MARGINS EVALUATION

APPENDIX B

COMPARISON BETWEEN A REDUCED–SCOPE
AND FULL–SCOPE SEISMIC MARGINS EVALUATION

There are differences between the reduced-scope and full-scope margins evaluation both in the extent of the systems analysis and in the amount of quantification of HCLPF values for equipment identified in the walkdown. The comparison is presented in Table B.1. The emphasis on walkdown and not on quantification also applies to the performance of containment and containment systems (that is containment performance analysis should concentrate on identifying seismically induced vulnerabilities and sequences different from those obtained from the IPE), USI A–45 (Decay Heat Removal Requirements), and GI–131 (In-Core Flux Mapping System).

B.1 Elements Preserved

The following elements of the seismic margins methodology must be preserved; that is, they must be identical in the reduced-scope and full-scope evaluation:

1. For either the NRC or EPRI methodology, the systems engineers must perform significant pre-walkdown work that should be preserved in a reduced-scope evaluation. In the NRC methodology, this involves defining initiating events, defining event trees and the safety functions involved, and identifying systems and components necessary to carry out these functions. In the EPRI methodology, this involves defining success paths (primary and alternative) and the systems and components involved in these paths. For both methodologies, the thrust of this work is to narrow the scope and focus the effort of the key element of the review, the *walkdown*.

2. For either the NRC or the EPRI methodology, the seismic capability evaluation engineers must perform significant pre-walkdown work that should be preserved in the reduced-scope evaluation. In each methodology, this involves developing an understanding of the seismic input to the plant and the seismic design basis and realistic ground and floor response spectra. It also involves pre-walkdown screening of the key systems and components identified by the systems engineers so as to make the walkdown itself most efficient. The thrust of the screening is to identify items thought to have very high HCLPF values, items suspected of having low HCLPF values, and therefore lists of items to be examined at various levels of detail during the walkdown.

3. The reduced-scope evaluation should be identical in quality and effort to that as for the full-scope margins methodology. One crucial feature is that it should involve interactions among seismic capability evaluation engineers, systems engineers, and the licensee's plant operations personnel. The walkdown team should visually inspect pertinent structures, equipment, and anchorages consistent with the full-scope NRC or EPRI methodology. If potentially vulnerable components are found during the walkdown, a capacity check may be necessary using the applicable SSE ground response spectra. These results should be documented. Data sheets similar to those found in Appendix I of EPRI NP–6041 should be used to document the walkdown. A review of construction drawings for structural details that can not be seen in the field should be performed.

4. While the post-walkdown assessment effort for a reduced-scope evaluation should be identical in quality to that in the full-scope margins methodology its thrust and level of effort are different because sequence-level (NRC) or success path-level (EPRI) HCLPFs will not be computed. Instead, its emphasis should be on identifying possible weak-link items that may need strengthening.

B.2 Reductions

The following, although needed in the full-scope margins methodology, are not needed in a reduced-scope margins evaluation

B.2.1 NRC Seismic Margins Methodology

1. The systems engineers need not prepare or quantify fault trees and Boolean expressions representing accident sequences. Also, since fault trees will not be developed, these engineers need not combine nonseismic failure basic events with seismically initiated failures in any rigorous fashion, although the existence of those non-seismic failures, if identified, should be noted and their importance assessed in the course of the margin evaluation.

2. The seismic capability evaluation engineers need not develop HCLPF capacity values for all of the key equipment items that would be represented on the sequence level Booleans (which will not be developed). It follows that they can not develop a plant-level HCLPF capacity value.

B.2.2 EPRI Seismic Margins Methodology

The seismic capability evaluation engineers need not develop HCLPF capacity values for all of the key equipment items found on the success paths (primary and alternative) being studied. It follows that they can not develop any success-path-level HCLPF capacity values that would be taken as representing the plant-level HCLPF capacity.

Table B.1 Reduced–Scope Margins Method

Based on NRC Seismic Margins Methodology (NUREG/CR–4482, Chapter 4)

Step No.	Description	In Reduced Program?
1	Selection of Earthquake Review Level	Not applicable, NRC designates sites that qualify
2	Initial Systems Review	Yes, in entirety
3	Initial Component HCLPF Categorization	Yes, in entirety
4	First Plant Walkdown	Yes, in entirety
5	Systems Modeling Finalize Event Trees: Fault Tree Development:	Yes No
6	Second Plant Walkdown	Only as needed
7	Systems Model Development	No
8	Margin Evaluation of Components, Plant	No

Based on EPRI Seismic Margins Methodology (EPRI NP–6041, Chapter 2)

Step No.	Description	In Reduced Program?
1	Selection of Seismic Margins Earthquake	Not applicable, NRC designates sites that qualify
2	Selection of Assessment Team	Yes, in entirety
3	Pre-Walkdown Preparation Work	Yes, in entirety
4	Systems and Element Selection Walkdown	Yes, in entirety
5	Seismic Capacity Walkdown	Yes, in entirety
6	Subsequent Walkdowns	Only as needed
7	Seismic Margin Assessment Work	No

APPENDIX C

DETAILED DOCUMENTATION AND REPORTING GUIDELINES

APPENDIX C
DETAILED DOCUMENTATION AND REPORTING GUIDELINES

This appendix provides the guidelines for detailed documentation and reporting format and content for the IPEEE submittals. The major parts of this appendix are the guidelines for seismic analysis (Section C.2), internal fire analysis (Section C.3), other analyses (Section C.4), specific safety features and plant improvements (Section C.1.4), and the licensee review team (Section C.1.5). The licensees are requested to submit their IPEEE reports using the standard table of contents given in Table C.1 or provide a cross reference. This will facilitate review by the NRC and promote consistency among various submittals. The contents of the elements of this table are discussed in sections below.

The level of detail needed in the documentation should be sufficient to enable NRC to understand and determine the validity of all input data and calculation models used, to assess the sensitivity of the results to all key aspects of the analysis, and to audit any calculation. It is not necessary to submit all the documentation needed for such an NRC review. Relevant documentation should be cited in the IPEEE submittal, and be available in easily retrievable form. The guideline for judging the adequacy of retained documentation is that independent expert analysts should be able to reproduce any portion of the results of the calculations in a straight forward, unambiguous manner. To the extent possible, the retained documentation should be organized along the lines identified in the areas of review. Any information that is comparable to that provided under the IPE for internal events can be incorporated by reference.

C.1 General

C.1.1 Conformance with Generic Letter and Supporting Material

Certification that an IPEEE has been completed and documented as requested by Generic Letter 88-20, Supplement 4. The certification should also identify the measures taken to ensure the technical adequacy of the IPEEE and the validation of results, including any uncertainty, sensitivity, and importance analyses.

C.1.2 General Methodology

Provide an overview description of the methodology employed in the IPEEE for each external event.

C.1.3 Information Assembly

Reporting guidelines include:

1. Plant layout and containment building information not contained in the Final Safety Analysis Report (FSAR).

2. A concise description of plant documentation used in the IPEEE, (e.g., the FSAR; system descriptions, procedures, and licensee event reports); and a concise discussion of the process used to confirm that the IPEEE represents the as-built, as-operated plant. The intent of such a confirmation is not to propose new design reverification efforts on the part of the licensees but to account for the impact of previous plant modifications or modifications conducted within the IPEEE framework.

3. A description of the coordination activities of the IPEEE teams among the external events (e.g., for seismically induced fires).

C.1.4 Submittal of Specific Safety Features and Potential Plant Improvements

The licensee should provide a discussion of the criteria used to define vulnerabilities for each external event evaluated. The licensee should list any potential improvements (including equipment changes as well as changes in maintenance, operating and emergency procedures, surveillance, staffing, and training programs) that have been selected for implementation based on the IPEEE (a schedule for implementation should be provided) or that have already been implemented. A discussion of anticipated benefits in terms of averted potential risk or increased plant seismic capacity as well as drawbacks to any improvements should be provided. Those improvements that have been taken credit for in the analysis and have not yet been implemented at the plant, should be specifically highlighted in the submittal.

C.1.5 IPEEE Team and Peer Review

The basis for requiring the involvement of the licensee's staff in the IPEEE review is the belief that the maximum benefit from the performance of an IPEEE would be realized if the licensee's staff were involved in all aspects of the examination and that involvement would facilitate integration of the knowledge gained from the examination into operating procedures and training programs. Thus the submittal should describe licensee staff participation and the extent to which the licensee was involved in all aspects of the program.

The submittal should also contain a description of the peer review performed, the same type of review as requested for the internal event IPE, the results of the review team's evaluation, and a list of the review team members.

C.2 Seismic Events

Section C.2.1 describes submittal information guidelines for licensees who choose the seismic PRA for seismic IPEEE, while section C.2.2 describes information guidelines for licensees who choose the seismic margin method for the seismic IPEEE. The submittal should be presented in conformance with Table C.1.

C.2.1 Seismic PRA Methodology

The following information on the seismic IPEEE should be documented and submitted to the NRC:

1. A description of the methodology and key assumptions used in performing the seismic IPEEE.

2. The hazard curve(s) (or table of hazard values) used and the associated spectral shape used in the analysis. Also, if an upper bound cutoff to ground motion of less than 1.5g peak ground acceleration is assumed, the results of sensitivity studies to determine whether the cutoff affected the overall results and the delineation and ranking of seismic sequences.

3. A summary of the walkdown findings and a concise description of the walkdown team and the procedures used.

4. All functional/systemic seismic event trees as well as data (including origin and method of analysis). Address to what extent the recommended enhancements have been incorporated in the IPEEE. A description of how nonseismic failures, human actions, dependencies, relay chatter, soil liquefaction, and seismically induced floods/fires are accounted for. Also, a list of important nonseismic failures with a rationale for the assumed failure rate given a seismic event.

5. A description of dominant functional/systemic sequences leading to core damage along with their frequencies and percentage contribution to overall seismic core damage frequencies (for both LLNL and EPRI hazard curves if used). Sequence selection criteria are provided in GL 88–20 and NUREG-1335. If either hazard curve causes a sequence to meet these criteria, that sequence should be included. The description of the sequences should include a discussion of specific assumptions and human recovery actions.

6. The estimated core damage frequency (for both the LLNL and EPRI hazard curves, if used) and plant damage state frequencies, the timing of the core

damage, including a qualitative discussion of uncertainties and how they might affect the final results, and contributions of different ground motions to core damage frequencies.

7. Any seismically induced containment failures and other containment performance insights. Particularly, vulnerabilities found in the systems/functions which will lead to early containment failure that might result in high consequences. This includes: isolation, bypass, integrity, and systems (e.g., igniters) required to prevent early failure. The computed fragilities of containment components, systems, and functions as applicable should be provided. The licensee may submit computed HCLPFs associated with containment performance (Optional).

8. A table of fragilities, both generic and plant-specific, used for screening as well as in the quantification. The estimated fragilities for the plant, dominant sequences, and dominant components should be reported. (Optional: The estimated HCLPF for the plant, dominant sequences, and components with and without nonseismic failures and human actions may be submitted by the licensee.)

9. Documentation with regard to other seismic issues (Section 6) addressed by the submittal, the basis and assumptions used to address these issues, and a discussion of the findings and conclusions. Evaluation results and potential improvements associated with the decay heat removal function and movable incore flux mapping system (for Westinghouse plants) should be specifically highlighted.

10. A Discussion of nonseismic failures and human actions that are significant contributors, or have impacts on results.

11. When an existing PRA is used to address the seismic IPEEE, the licensee should describe sensitivity studies related to the use of the initial hazard curves, supplemental plant walkdown results and subsequent evaluations, and relay-chatter evaluations. The licensee should examine the above list to fill in those items missed in the existing seismic PRA (See Section 3.1.2).

C.2.2 Seismic Margins Methodology

The following information on the seismic IPEEE should be documented and submitted to the NRC for a full-scope or a focused–scope SMM review:

1. A description of the methodology and a list of important assumptions, including their basis, used in performing the seismic IPEEE. Address the extent to which the following were taken into account:

nonseismic failures, human actions, dependencies, relay chatter, soil liquefaction, and seismically induced floods/fires. Also, a list of important nonseismic failures with a rationale for the assumed failure rate given a seismic event.

2. A summary of the walkdown results and a concise description of the walkdown team and procedures used.

3. All functional/systemic seismic event trees data (including origin and method of analysis) when NRC SMM is used.

4. A description of the most important sequences and important minimal cutsets (for both seismic and nonseismic failures) leading to core damage (NRC method) or a description of the success paths and procedures used for their selection and of each component in the controlling success path (EPRI method).

5. Any seismically induced containment failures and other containment performance insights. Particularly, vulnerabilities found in the systems/functions which will lead to early containment failure and high consequences. This includes: isolation, bypass, containment integrity and systems (e.g., igniters) required to prevent early failure. Also, computed fragilities (if used) and HCLPFs of containment components, systems, and functions as applicable.

6. A table of fragilities (if used) and HCLPFs, both generic and plant-specific, used for screening as well as in the quantification. The estimated fragilities (if used) and HCLPFs for the plant, dominant sequences, and dominant components should be reported.

7. Documentation with regard to other seismic issues (Section 6) addressed by the submittal, the basis and assumptions used to address these issues, and a discussion of the findings and conclusions. Evaluation results and potential improvements associated with the decay heat removal function and movable incore flux mapping system (for Westinghouse plants) should be specifically highlighted.

8. For the NRC method, provide a discussion of nonseismic failures and human actions that are significant contributors to or have an impact on the results.

The following is the information that should be documented and submitted to the NRC for a reduced-scope SMM review:

1. A description of the procedures used to identify systems and components for the walkdown in performing the seismic IPEEE.

2. A summary of the walkdown findings and a concise description of the walkdown team and procedures used.

3. A discussion and the results of any specific component capacity evaluations performed, the methods used, and assumptions.

4. Documentation with regard to other seismic issues (Section 6) addressed by the submittal, the basis and assumptions used to address these issues, and a discussion of the findings and conclusions. Evaluation results and potential improvements associated with the decay heat removal function and movable incore flux mapping system (for Westinghouse plants) should be specifically highlighted.

C.3 Internal Fires

The information on the internal fires IPEEE identified below should be documented and submitted to the NRC.

1. A description of the methodology and key assumptions used in performing the fire IPEEE and a discussion of the status of Appendix R modifications.

2. A summary of the walkdown findings and a concise description of the walkdown team and the procedures used. This should include a description of the efforts to ensure that cable routing used in the analysis represents as-built information and the treatment of any existing dependence between remote shutdown and control room circuitry.

3. A discussion of the criteria used to identify critical fire areas and a list of critical areas, including (a) single area in which equipment failures represent a serious erosion of safety margin, and (b) same as (a), but for double or multiple areas sharing common barriers, penetration seals, HVAC ducting, etc.

4. A discussion of the criteria used for fire size and duration and the treatment of cross-zone fire spread and associated major assumptions.

5. A discussion of the fire initiation data base, including the plant-specific data base used. Provide documentation in each case where the plant-specific data used is less conservative than the data base used in the approved fire vulnerability methodologies. Describe data handling method, including major assumptions, the role of expert judgment, and the identification and evaluation of sources of data uncertainties.

6. A discussion of the treatment of fire growth and spread, the spread of hot gases and smoke, and the analysis of detection and suppression and their associated assumptions, including the treatment of suppression-induced damage to equipment.

7. A discussion of fire damage modeling, including the definition of fire-induced failures related to fire barriers and control systems and fire induced damage to cabinets. A discussion of how human intervention is treated and how fire-induced and non-fire-induced failures are combined. Identify recovery actions and types of fire mitigating actions taken credit for in these sequences.

8. Discuss the treatment of detection and suppression, including fire fighting procedures, fire brigade training and adequacy of existing fire brigade equipment, and treatment of access routes versus existing barriers.

9. All functional/systemic event trees associated with fire-initiated sequences.

10. A description of dominant functional/systemic sequences leading to core damage along with their frequencies and percentage contribution to overall core damage frequencies due to fire. Sequence selection criteria are provided in GL 88–20 and NUREG–1335. The description of the sequences should include a discussion of specific assumptions and human recovery action.

11. The estimated core damage frequency, the timing of the associated core damage, a list of analytical assumptions including their bases, and the sources of uncertainties.

12. Any fire induced containment failures identified as being different from those identified in the internal events analysis.

13. Documentation with regard to the decay heat removal function and Fire Risk Scoping Study issues addressed by the submittal, the basis and assumptions used to address these issues, and a discussion of the findings and conclusions. Evaluation results and potential improvements should be specifically highlighted.

14. When an existing PRA is used to address the fire IPEEE, the licensee should describe sensitivity studies related to the use of the initial hazard, supplemental plant walkdown results and subsequent evaluations. The licensee should examine the above list to fill in those items missed in the existing fire PRA.

C.4 High Winds, Floods, and Others

The following information on the high winds, floods, and others portion of the IPEEE should be documented and submitted to the NRC:

1. A description of the methodologies used in the examination.

2. Information on plant–specific hazard data and licensing bases.

3. Identified significant changes not reported per 10 CFR 50.71(e) (See Section 5.2.2), if any, since OL issuance with respect to high winds, floods, and other external events.

4. Results of plant/facility design review to determine their robustness in relation to NRC's 1975 SRP criteria.

5. Results of the assessment of the hazard frequency and the associated conditional core damage frequency if step 4 of Figure 5.1 is used.

6. Results of the bounding analysis if step 5 of Figure 5.1 is used.

7. All functional event trees, including origin and method of analysis (PRA only).

8. A description of each functional sequence selected, including discussion of specific assumptions and human recovery action (PRA only).

9. The estimated core damage frequency, the timing of the associated core damage, a list of analytical assumptions including their bases, and the sources of uncertainties, if applicable (PRA only).

10. A certification that no other plant-unique external event is known that poses any significant threat of severe accident within the context of the screening approach for "High Winds, Floods, and Others."

Table C.1 Standard Table of Contents for IPEEE Submittal

1. Executive Summary

   1.1 Background and Objectives
   1.2 Plant Familiarization
   1.3 Overall Methodology
   1.4 Summary of Major Findings

2. Examination Description

   2.1 Introduction
   2.2 Conformance with Generic Letter and Supporting Material
   2.3 General Methodology
   2.4 Information Assembly

3. Seismic Analysis

   3.0 Methodology Selection
   3.1a Seismic PRA
       3.1.1 Hazard Analysis
       3.1.2 Review of Plant Information and Walkdown
       3.1.3 Analysis of Plant System and Structure Response
       3.1.4 Evaluation of Component Fragilities and Failure Modes
       3.1.5 Analysis of Plant Systems and Sequences
       3.1.6 Analysis of Containment Performance
   3.1b Seismic Margins Method (SMM) (NRC, EPRI, or Reduced SMM)
       3.1.1 Review of Plant Information, Screening, and Walkdown
       3.1.2 System Analysis
       3.1.3 Analysis of Structure Response
       3.1.4 Evaluation of Seismic Capacities of Components and Plant
       3.1.5 Analysis of Containment Performance
   3.2 USI A-45, GI-131, and Other Seismic Safety Issues

4. Internal Fires Analysis

   4.0 Methodology Selection
   4.1 Fire Hazard Analysis
   4.2 Review of Plant Information and Walkdown
   4.3 Fire Growth and Propagation
   4.4 Evaluation of Component Fragilities and Failure modes
   4.5 Fire Detection and Suppression
   4.6 Analysis of Plant Systems, Sequences, and Plant Response
   4.7 Analysis of Containment Performance (If Applicable)
   4.8 Treatment of Fire Risk Scoping Study Issues
   4.9 USI A-45 and other Safety Issues

5. High Winds, Floods, and Others

   5.1 High Winds
   5.2 Floods
   5.3 Transportation and Nearby Facility Accidents
   5.4 Others

6. Licensee Participation and Internal Review Team

   6.1 IPEEE Program Organization
   6.2 Composition of Independent Review Team
   6.3 Areas of Review and Major Comments
   6.4 Resolution of Comments

7. Plant Improvements and Unique Safety Features

8. Summary and Conclusions (including proposed resolution of USIs and GIs)

APPENDIX D

NRC RESPONSE TO COMMENTS AND QUESTIONS

APPENDIX D
NRC RESPONSE TO COMMENTS AND QUESTIONS

D.1 Introduction and Summary

The NRC staff conducted an IPEEE Workshop on September 11-13, 1990 at the Pittsburgh Hilton in Pittsburgh, Pennsylvania. The objectives of the workshop were to discuss the IPEEE process and to solicit questions and comments on the guidance for performing the IPEEE and for reporting the results of the review. The schedule of the IPEEE Workshop was announced in the Federal Register (55 FR 30332) July 25, 1990, and a preliminary agenda of the workshop was published on August 10, 1990, in the *Federal Register* (55 FR 32712).

All the questions and comments raised at the workshop or submitted after the workshop were categorized into several major subject areas. This appendix summarizes these questions and comments and the NRC staff responses (SR) to them. The most significant comments, concerns, and questions, together with staff response (SR), are summarized below. This summary also serves to highlight the major changes made in going from the proposed to the final documents.

1) *Backfit Analysis: Should a regulatory backfit analysis of the proposed IPEEE effort be performed prior to issuance of the IPEEE generic letter?*

SR: The staff does not believe, as a legal matter, that a 10 CFR 50.109 type backfit analysis is needed for the IPEEE generic letter (memorandum from W. Parler to Commissioners, dated July 27, 1990). Because the request to perform the IPEEE is considered to be a request under 10 CFR 50.54(f), the staff has performed a 50.54(f) analysis, which is included as Appendix 5 to Supplement 4 of Generic Letter 88-20. However, in view of the significant licensee resource commitment required to respond to this information request and in the interest of a prudent policy, the staff has completed a value-impact analysis and has included it as an attachment to this appendix. This analysis shows that, based on previous experience with the evaluation of severe accidents initiated by external events, the IPEEE has the potential to identify items that, if corrected, would result in substantial increases in safety, and that the cost of corrections, including the cost of the IPEEE, would be commensurate with averted potential risk to severe accidents.

2) *Cost Estimates and Resource Requirements: The overall cost of the IPEEE was underestimated.*

SR: The staff derived its estimates of cost and resource requirements to perform an IPEEE from the actual costs spent on two NUREG-1150 plants and the cost spent on the Hatch seismic review extrapolated to the IPEEE scope. At the IPEEE workshop, certain industry estimates were presented that were either less than or comparable to the staff's estimates. The staff recognizes that there are uncertainties in the costs because there are uncertainties associated with the analysis of external events. However, the staff believes that there are ways to keep the cost under control. If additional questions arise regarding the IPEEE process and the associated guidelines, the staff will meet with licensees and respond to those questions.

3) *Schedule and Resource Availability: Schedule and resource availability for performing the IPEEE are of concern.*

SR: The methods identified in this report are not new and have been used and discussed extensively in the past. Probabilistic risk assessment (PRA) procedures for assessing the risk associated with external events have been used since the late 1970's. The NRC seismic margins method was published in 1985; the Electric Power Research Institute (EPRI) seismic margins method was published in 1988. These methods were derived from the insights gained from available seismic PRAs. They were widely discussed at many conferences and workshops and were used at three plant sites. Procedures for the seismic walkdown, one of the most important ingredients in the seismic IPEEE, are similar to those that will be used in the implementation of Unresolved Safety Issue (USI) A-46, "Verification of Seismic Adequacy of Equipment in Operating Plants." A number of trial walkdown training workshops with a number of participants from the utilities were conducted in the past by the Seismic Qualification Utility Group (SQUG), which developed the walkdown procedures for USI A-46. Finally, the event trees and fault trees developed for the internal event IPE, which was initiated about 18 months before the IPEEE, will be available for use in the IPEEE. Therefore, the staff believes that a large pool of talent is available (as evidenced by the number [approximately 25] of consultants and consulting firms represented at the IPEEE workshop), and that within the 3-year period to perform the IPEEE, licensees can develop or obtain the necessary expertise to conduct the IPEEE. However, as with the internal events IPE, the staff will consider extending this date on a case-by-case basis.

4) *Licensee Response Time: The initial response time of 60 days to identify the methodologies for completing IPEEE is too short.*

SR: The staff believes that it is appropriate to extend the licensee's initial response time from 60 days to 180 days to allow for some essential preparatory work (i.e., the processing of bids, completion of the development of the alternative fire evaluation methodology by the Nuclear Management and Resource Council (NUMARC) and EPRI, and the staff's review and assessment of this methodology). One hundred eighty days was selected in consideration of the current schedule for NUMARC/EPRI to complete the development and verification of their alternate fire evaluation methodology, and for the subsequent staff review. These activities are expected to be completed in July 1991 (see Response 6.1).

5) *Inclusion of Issues: The inclusion of (a) lightning, (b) volcanic activities, and (c) Generic Issue (GI) 103, "Design for Probable Maximum Precipitation (PMP)" in the IPEEE was questioned.*

SR: Licensees need to confirm that lightning or volcanic activity is not a dominant contributor to severe-accident risk at their nuclear power plant sites. The determination should be based on plant-specific experience. The concern related to lightning (as pointed out by the ACRS) is that lightning strikes, in addition to causing loss-of-offsite power, may damage instrumentation and control systems. If this had happened before at a site, the staff would expect the IPEEE for that specific plant to address this concern. In regard to volcanic activity, only two sites would be affected. In either case, a simple discussion will be sufficient for those plants not affected by these events. For plants that may be affected, a success screening process, such as the one described in this report can be used.

With regard to GI-103, the NRC acknowledged the importance of this new PMP criterion in Generic Letter 89-22 by requiring that future plants be designed against it (i.e., design basis). For existing plants, the NRC recommended that licensees review the material contained in GL 89-22 to determine whether they believe additional action is necessary; however, licensees need not change their design bases. For the IPEEE, the staff believes that assessing the potential for a PMP to cause a severe accident is justified, since the National Weather Service PMP data are being applied to future designs. However, the staff has clarified it's recom-

mendation by limiting the assessment of the effects to onsite flooding and roof ponding.

6) *Fire Evaluation*

6.1) *Expeditious NRC review of the NUMARC/EPRI alternate fire-evaluation methodology was requested.*

SR: The staff had previously committed to review an alternate methodology being developed by NUMARC and EPRI for evaluating fires. Currently, the staff is reviewing a NUMARC document describing the methodology and is waiting for more information from EPRI and NUMARC on the results of demonstration applications of this methodology at two nuclear plants (the staff anticipates this will come in April 1991). Following receipt of this information, the staff plans to complete its review, including discussions with the ACRS and the NUMARC. As stated in Response 4, these activities are not expected to be completed until July 1991. Therefore, the staff plans to respond separately to the NUMARC/EPRI alternate methodology, so as not to delay issuing the IPEEE generic letter and guidance document. However, as discussed in Response 4, additional time was given to licensees, so that they have the results of the staff review before they commit to a fire evaluation methodology in their IPEEE submittal plans.

6.2) *GI-57, "Effects of Fire Protection System Actuation on Safety-Related Equipment:" The relationship of this GI to the IPEEE was questioned.*

SR: The effect of fire suppressants on safety equipment is one of the safety issues identified in the Fire Risk Scoping Study, (NUREG/CR-5088) and may be a significant contributor to risk. Accordingly, it was raised as a generic safety issue and was also included in the IPEEE. The staff expects that if a licensee discovers a significant vulnerability in this area through the IPEEE, the licensee would address the problem and not await the GI-57 resolution. During the walkdown, licensees can collect relevant information on whether actuated fire protection systems would spray safety-related equipment, and can institute some protective measures to prevent the safety equipment from being sprayed by fire suppressants. The additional effort to collect this information during the walkdown should not be a burden. However, the formal resolution of GI-57 does not have to be a part of the IPEEE.

7) *Seismic Events: Treatment of seismic events needs clarification.*

7.1) *The need to use both the LLNL and EPRI seismic hazard curves was questioned.*

SR: The staff considered the difference between these two curves in specifying the enhancements for the seismic margins methods and the seismic PRA. However, based on the available information to date, the staff is unable to dispute the merit of either curve and considers both of them to be valid. The staff also believes that the added cost of using two curves should not be burdensome, based on what was spent on two NUREG-1150 plants. A utility may choose to use only a single curve, provided the higher one of the two curves is chosen.

7.2) *The use of a site-specific seismic curve (in lieu of the LLNL and EPRI curves) was suggested.*

SR: The LLNL and EPRI curves are "site-specific" seismic curves. Each used its standard methodology and uniform interpretation of data bases to calculate the seismic hazards for power plant sites in the Eastern United States. The use of other site-specific seismic hazard curves is an acceptable option subject to review and acceptance by the staff. However, the staff believes that the cost associated with the development of new site-specific seismic hazard curves could be very high and time consuming.

7.3) *The use of plant design bases in the seismic binning process was suggested.*

SR: The staff investigated the potential of using the seismic design basis as a parameter for making the initial binning assignments. Because there was insufficient technical basis for its use, it was not used initially. However, when considered in conjunction with the seismic hazard, the use of the seismic design bases provided a means for reducing the scope of the 0.3g margins review. Specifically, plant sites in the 0.3g bin were assigned to a full-scope or a newly defined focused-scope category. The full-scope category is essentially the review specified in the draft generic letter and guidance document that were distributed for public comment, whereas the focused-scope review represents a reduced scope review. The primary purpose of this further subdivision is to reduce the level of review effort, mainly in the relay chatter area, for plants with a lower hazard or higher design basis. Plants with a relatively higher hazard and lower seismic design basis should perform a more detailed study than the other plants. (Grouping of plants into various categories is discussed in Section 3 and Appendix A of this report.) Of the 56 sites originally assigned to the 0.3g category, 7 remain in the full-scope category, the remainder moved to the focused-scope category.

7.4) *The scope of the relay chatter evaluation was questioned.*

SR: Detailed relay chatter studies conducted at the Hatch and Diablo Canyon plants showed that considerable resources were expended to perform the relay chatter review; and using existing procedures, operators could solve the relay chatter problems identified at these plants. However, the staff and industry consultants are concerned that such conclusions cannot be considered generic without some additional plant reviews.

Therefore, as discussed above, the staff is recommending that the 0.3 g bin be subdivided into two categories, a full-scope and a focused-scope category. For plants in the full-scope category, licensees will have to evaluate the relay chatter consistent with the approach discussed in EPRI NP-6041 or its equivalent. For reduced-scope review, the implementation of USI A-46 program will provide information for satisfying the IPEEE provisions. Note that licensees can perform the IPEEE with considerable efficiency, if they take advantage of the lessons learned from the Hatch and Diablo Canyon relay chatter evaluations. For plants in the focused-scope category, a lower level of effort is recommended; this would entail looking for and addressing low-capacity relays only.

D.2 Detailed Response to Public Comments and Questions

This section paraphrases, summarizes, and categorizes into subject areas, questions and comments either raised at the workshop or received by the staff (16 parties submitted written comments, see sources of comments). The NRC staff response is also provided. Table D.1 contains a listing of the subject areas discussed in this section. The workshop transcript and a copy of the comments received are available in the NRC Public Document Room.

Table D-1 Categorization of Question and Comments

1. IPEEE, IPE, CPI, and Accident Management
2. Backfit analysis
3. Cost estimates and resource requirements
4. Schedule and response time
5. High winds, floods, and transportation and nearby facility accidents
6. Internal fires
7. Seismic events

1. IPEEE, IPE, CPI, and Accident Management

1.1 How do the IPE, IPEEE, and the accident management all relate to the design bases of the plant in terms of identified plant vulnerabilities, improvements, and potential increase in risk? How do the plant operators make the day-to-day decisions when PRA insights and Tech Specs are in conflict? (Ref. D.16, p. 40)

SR: The thrust of the whole severe accident program is to reduce the likelihood of severe accidents and their consequences. As such, they are looking at accident scenarios beyond the traditional design basis envelope. If a vulnerability is identified and a fix is proposed, it is important to make sure that the proposed fix has no adverse effect on the plant. If a licensee makes modifications to the plant, which resulted in a change of the plant design basis, then that must be documented, tracked, and accounted for, in accordance with the provisions of 10 CFR 50.59. If the PRA identifies a conflict with the Technical Specifications or operating procedures, the licensee should examine the reason for the conflict. It is important that the licensee not make plant or procedural modifications without understanding the basis behind the PRA conclusions. For example, if the underlying model in the PRA was developed with simplified assumptions and modeling techniques, it might be prudent to perform a more realistic evaluation to assure that the modeling assumptions have not biased the results in an inappropriate manner.

1.2 After the utilities had factored vulnerabilities identified through IPE and IPEEE in their emergency operating procedures (EOPs) and the Tech Specs that support the EOPs, what else would the utilities be required to do for the accident management? Any specific example of a guideline that the staff would be putting forth as part of accident management? (Ref. D.16, p. 46)

SR: It is important to recognize that accident management responses are not just limited to emergency procedure guidelines (EPGs) or EOPs. Technical support and the kind of guidance and personnel training that are needed should be part of accident management. IPE and IPEEE results should be considered as an information source which provides inputs to training programs and to the development of emergency preparedness exercises. Accident management takes that information and uses it in the planning, training, exercises, and to establish the communication and the feedback mechanism at the utilities. Specific guidance and examples are still being developed by industry (NUMARC) and NRC,

and will be addressed in a future generic letter on accident management.

1.3 The Severe Accident Policy Statement is silent on external events, thus, there is no need to do much beyond what is already done for internal event IPEs. (Ref. D.16, p. 53)

SR: The Commission Policy Statement identified the need to seek vulnerabilities systematically at all operating plants. It didn't distinguish between internal and external events. However, PRA studies have shown that external events, in particular seismic and fire, are principal contributors to overall risk. Accordingly, the staff recommended to the Commission in SECY-86-162, dated May 22, 1986, that external events be included in implementation of the Severe Accident Policy Statement.

1.4 IPEEE, basically, is an evaluation looking at a point in time, a snapshot in time, is there an intent of keeping it living? (Ref. D.16, p. 78)

SR: The staff is treating the IPEEE as a one-time evaluation. There is no requirement to keep it living. However, based on utilities' experience, once one has gone through the process and invested the resources and constructed the PRA or equivalent, it would be useful to keep it up to date. However, it is really up to the utility whether or not to keep it living.

1.5 Is it correct to assume that there is no requirement for the pedigree of the program, that it can be basically be a study without a QA type of pedigree? (Ref. D.16, p. 80)

SR: There is no requirement for an Appendix B-type QA program to check the IPE or IPEEE. The licensees should perform an ongoing internal quality assurance effort to ensure that the results of the IPE and IPEEE are factual and represent the as-built, as-operated plant. Typically, licensees will define "pinch points" in their performance of a risk assessment to stop and assess the progress and quality of their effort to date. As in the internal event IPE, the staff is asking for a peer review as part of the IPEEE. The peer review provides a type of QA function.

1.6 In the area of other external events that are not included in the IPEEE, does the utility need to develop a hazard curve associated with that particular event? (Ref. D.16, p. 81)

SR: No. The staff is not asking the utilities to justify not including those events. However, if a utility knows of a particular hazard that is greater than what the staff has considered in the generic study, the utility should consider including it in the IPEEE.

1.7 Is sabotage included in the IPEEE? (Ref. D.16, p. 82)

SR: Sabotage is not included.

1.8 Florida Power & Light has a policy that upon approaching hurricanes in S. Florida, the unit will be shut down and the unit will go to a Mode 3 or a Mode 4 in advance of the hurricane, does that mean the hurricane need not be considered in the IPEEE? (Ref. D.16, p. 82)

SR: In general, the shutdown mode is outside of the IPEEE scope. However, in a case like this, the licensee should make sure that the plant can be shutdown and maintained in a safe shutdown condition (USI A–45 requirements). In other words, the combined frequency of the hurricane and failure to shutdown and to maintain the plant in a safe shutdown condition needs to be assessed.

1.9 Faced with large uncertainties, how are risk, human reliability, operational reliability, maintenance reliability, etc. associated with external events to be quantified? (Ref. D.16, p. 84)

SR: The staff recognizes that there are significant uncertainties in quantifying risks associated with both internal and external events. Thus, the staff has deemphasized the bottomline numbers in both the IPE, and more importantly, for the IPEEE.

1.10 When should the improvements resulting from the IPE process be carried out, right after the IPE or wait until the IPEEE is completed? (Ref. D.16, p. 87)

SR: The staff has looked at the interaction between the internal and external events. The staff has concluded that it is unlikely that the cost effective improvements based on internal IPEs would have a negative impact on safety for external events (see Section 6.3.1.2). The Generic Letter states that it expects each licensee to move expeditiously to correct any vulnerabilities that it determines warrants correction.

1.11 Since the purpose of the IPEEE is to gain a qualitative understanding of core damage frequency, not quantitative, is it good enough for seismic PRAs to just report seismic risk in terms of high, medium, or low, instead of putting in numbers? (Ref. D.16, p. 89)

SR: No. Core damage frequencies are requested from the seismic PRA so as to obtain insights and the relative ranking of the accident sequences, and

dominant components as well as assess the findings against the reporting criteria.

1.12 In the draft generic letter, how will this information be used is rather general. It seems that it is up to the utilities to determine what to do with the results, how they see fit; make the change or justify not making the change. However, in the generic letter it also says that the NRC will assess whether the conclusions the licensee draws from the IPEEE regarding changes to the plant systems or components are adequate. May be you should include some kind of an example of either a positive or a negative finding that you've made that can provide some sort of guide as to what is adequate and what is not adequate. (Ref. D.16, p. 92)

SR: If the staff disagrees with what a licensee did, the options that are open to the staff are contained in the regulations. The staff may request additional information via questions pertaining to the submittal, or may impose plant modifications via the backfit rule.

1.13 Vulnerabilities need to be tied to core damage risk. If you can't tie a vulnerability to a core damage risk, then it's not a vulnerability. The Severe Accident Policy Statement says that if you identified a vulnerability, then what you do next is to see if by fixing that vulnerability you can reduce the risk of core damage. So it seems that you have to have a quantitative number for core damage before and after you fix the identified vulnerability. (Ref. D.16, p. 94)

There are several stages in the process: (1) To identify a vulnerability; (2) To identify fixes for that vulnerability; (3) To determine if it's substantial; and (4) To determine if it's cost-beneficial. The cost-beneficial side of it has to be determined by the reduction in risk to the population outside the plant, which, in effect, requires a containment failure.

So we have to identify the vulnerability list and then follow these steps. Because the IPEEE implements severe accident policy, we all have to make sure that we're implementing it in the way intended.

SR: The process described is what the staff would go through if we chose to backfit a plant. It's not necessarily what the utilities might do in their plant. We leave it up to the utilities to decide what process they would use and how they would define a vulnerability.

1.14 Clarify the IPEEE Objectives. (Ref. D.1)

SR: The purpose of the IPEEE is to gain a qualitative understanding of core damage frequency, not quantitative. Section 1.2 provides the discussion of this aspect and also points out that some methods have been developed for evaluating external event hazard and identifying vulnerabilities that do not produce estimates of damage frequency. Seismic margins method is cited as a specific example. The objectives were reworded to emphasize this point.

2. Backfit Analysis

2.1 The staff has stated that the Office of the General Counsel is looking into whether the request for an IPEEE should be under 50.54(f) or the backfit rule. If the Office of the General Counsel does determine that the backfit rule applies, would it be correct to assume that this supplement to the generic letter will be issued similar to Generic Letter 89-16, the Hardened Vent of the Mark I's, where doing the action or performing the IPEEE would be voluntary, and for those utilities who do not volunteer to perform it, plant-specific backfit analyses would be performed using plant-specific values and criteria in terms of our own resources required and the scope of the analyses that an individual plant would have to perform? (Refs. D.13 and D.16, pp. 90–91)

SR: The staff has determined that a backfit analysis is not needed for the IPEEE.

2.2 If the General Counsel does determine that this can go ahead under a 50.54(f) request for information, will the utilities have an opportunity to look at that for themselves and perhaps appeal that decision or do their own analyses, whether we feel that backfit rule actually applies or not? (Ref. D.16, p. 92)

SR: All utilities will be required to respond to the IPEEE 50.54(f) request.

2.3 Without some kind of numerical or specific figure of merit, how can one really say that it's cost-effective to implement one type of fix over another, or even to make a fix at all? (Ref. D.16, p. 92)

SR: The staff traditionally uses a cost benefit figure of merit of $1000/man-rem in imposing new requirements.

That does not constrain a licensee to use the same cost benefit as their criterion for what to fix or what not to fix. Any criterion that the licensee chooses to use should be justified.

2.4 What are the criteria that NRC will use in determining what to fix? (Ref. D.16, p. 154)

SR: If the staff determines that a fix is required to bring the plant into conformance with the regulations, no cost-benefit or other analysis is required, and the staff can seek to have the fix implemented, if necessary by Order. If the staff determines that a fix desired by the staff is beyond the design bases or Regulations of the Commission, the staff must prepare a backfit analysis and submit it to the Commission. A backfit analysis does not have to include a probabilistic risk evaluation, but can instead be presented relying primarily on engineering judgment.

2.5 The Severe Accident Policy Statement doesn't give you any probabilistic numbers that you can really work with. However, in June 1990, the Commission directed the staff to consider 10^{-4} core melt frequency as safe enough as in the Safety Goal Policy Statement.

We propose to link "how safe is safe enough" with the Safety Goal Policy Statement about undue risk in severe accident with adequate protection in the backfit rule. We could make a very good case that plants currently are safe enough especially if we use that number on the individual plant level. (Ref. D.16, p. 156)

SR: The Commission Safety Goal is not just 10^{-4} per year for core melt. The Commission safety goals are the quantitative health objectives for fatalities. The staff had recommended that subsidiary objectives (e.g., 10^{-4} CDF) be established as a way to implement the Safety Goals in a practical manner.

In addition, the Safety Goals are not to be used to judge individual plants. They are to be used to judge the acceptability of the NRC regulations.

2.6 There is a concern about the closure process, particularly in relation to the fact that both the EPRI and the Livermore hazard curves will be used in a seismic PRA.

What's going to happen is ultimately, if a seismic vulnerability would be expensive to repair, you are forced into doing some probabilistic type cost benefit analysis. Even though you may have done a seismic margins assessment to identify that vulnerability, I see no way that you are able to avoid not reverting to using the Livermore hazard curves in your ultimate decision-making process. I think this is the time that there ought to be some effort to resolve the difference between these two curves. (Ref. D.16, pp. 162–164)

SR: When we do backfit analysis and regulatory analysis, it's not uncommon to have areas of great uncertainty, even as large as the difference between these two curves. The EPRI and LLNL hazard curves display a level of that uncertainty. Some backfit evaluations are evaluated without having core damage/ risk values, but rather rely on engineering judgment.

The Commission makes use of all the information available in making its decisions. HCLPFs may also be an important consideration in making decisions regarding backfit.

If a licensee prefers to use a single curve in the seismic review, that is also acceptable provided the higher of the two curves is chosen. See response 7.8.

3. Cost Estimates and Resource Requirements

3.1 The staff's resource estimate for IPEEE is low. (Refs. D.9, D.13, and D.16, p. 42)

SR: The staff believes that the cost estimate for IPEEEs was developed conservatively. Obviously, there are uncertainties in the costs because there are uncertainties in how a licensee will choose to analyze external events. The staff has used previous utility and NRC experience in conducting external event analysis in arriving at the cost estimate and has attempted to clarify the scope of the IPEEE where questions on scope were raised. However, where questions still exist, licensees should come in and talk to the staff to make sure that they have a clear understanding of the IPEEE process.

3.2 Six person years was the estimate for the IPEEE cost or manpower resource requirement, what is the basis? (Ref. D.16, p. 155)

SR: The basis is discussed in Appendix 5 to Generic Letter 88-20, Supplement 4. Basically what the staff did was to estimate the overall effort required to perform the IPEEE and to use the cost spent for NUREG-1150 plants and the Hatch seismic review to estimate the IPEEE cost. We also received estimates from a few PRA companies. We do recognize that there are some costs that licensees will have to spend over and above just what it will cost the PRA company. We do believe that the staff's estimates are reasonable. K. Fleming of PLG Inc., whose firm has done most industry-sponsored PRAs, and D. Dube of NE Utilities provided estimates that are either less than or comparable to the staff's estimates. Table D.2 provides a comparison of the staff's and the industry resource estimates.

4. Schedule and Response Time

4.1 What is the schedule for IPEEE, staff review, and fixes? (Ref. D.16, p. 50)

SR: The staff intends to complete closure of severe accident issues in 1995. Accordingly, a three year completion period for the IPEEE is specified so as to give the staff time to review the submittals.

5. High Winds, Floods, and Transportation and Nearby Facility Accidents

5.1 The flooding criterion screens at a frequency of 10^{-5} per year, but the wind screens at 10^{-6} per year, why is there an inconsistency between flooding and wind? (Ref. D.16, pp. 144 & 150)

SR: The screening criteria are essentially a reporting criteria, which are consistent for all external events and internal events.

We're not using 10^{-5} per year flood frequency alone to screen out floods. Based on a number of flood studies, the judgment was made that the probable maximum flood has a 10^{-5}/yr. or less frequency. But that alone does not screen out the flood concern. Judgment was made that the conditional failure probability for a plant designed against the probable maximum flood is somewhere around 10^{-1}, so that the combined probability is 10^{-6} or less per year. A similar approach was used for the wind, where the design basis wind was usually selected to have a probability of less than 10^{-6} per year.

5.2 The item 2 on the flowchart for screening external events, identifying significant changes, does that refer to the hazard at the site or does that refer to the design of a plant? (Ref. D.16, p. 147)

SR: It refers to the hazard on the site and the land use of the general vicinity, for example, a new airport built nearby.

5.3 For item 3, review plant against the current standard review plan criteria, why do older plants need to do that? Obviously, some older plants were not designed using current methods and codes. (Ref. D.16, p. 147)

SR: Item 3 is a screening criterion only. If you know already that your plant does not meet the 1975 Standard Review Plan (SRP) criteria, you should move on to the next step in the evaluation process.

5.4 The inclusion of lightning and volcanic activity is questioned. (Refs. D.1, 8, & 11)

SR: The confirmation that lightning or volcanic activity is not a significant contributor to severe accidents at the nuclear power plant, should be assessed to the satisfaction of the utilities. A relatively simple discussion by the licensee to state why a licensee believes these issues to be unimportant, will be sufficient for these issues for most plants.

Table D.2 IPEEE Resource Estimates (Manhours)

Task	NUMARC[2]	Hatch[1] Actual	Today*	Future SMM+	PLG SPRA[3]
1. Select SME	0	200	200	0–200	
2. Select success paths	600	850	700	600–950	
3. Mod. seismic bld'g model	600	430	430	0–950	
4. Perform SSI & Dev. FRS	2000	1020	1020	0–2000	
5. Soil liq. evaluation	1000	500	500	0–1000	
6. Relay chatter evaluation	2000	2800	1700	1400–2100	
7. Pre-walkdown	160	160	100–200		
8. Walkdown preparation	200	200	100–300		
9. Seismic walkdown	900	900	900	600–1400	
10. SMM of outliers	1800	2680	2000	1200–2700	
11. Report & documentation	700	1320	1000	980–1200	
12. Walkdown travel expense	500				
13. Containment review isolation, bypass, etc	800				
Long term mitigation	400				
14. Misc. cost (startup, plant support, training, peer review, NRC interaction)	1200				
Total Seismic IPEEE	13000	11060	8810	4980–13000	1700–2700

	Surry	Peach Btm	PRA	PLG
Seismic event	1400	1320	1500–3000	1100–1800
Internal fires	350	450	750	900–1500
High winds	–	–	0–200	
External floods	150	250	0–300	
Others	–	–	0–650	
QA	450	450	200–500	
Plant support	1200	1300	1300–2500	1000–2000

Sources:
[1]* + D. P. Moore of Southern Company Services, Inc.
[2]NUMARC cost estimate of full-scope SMM of Nov. 13, 1990.
[3]PLG letter to NRC, dated Oct. 9, 1990

5.5 The requirement for assessing GI 103, Design for Probable Maximum Precipitation (PMP) and lack of specific guidance in the generic letter are questioned. (Refs. D 1, 8, 11, & 13)

SR: IPEEE imposes no requirements. With regard to the PMP, NRC acknowledged the importance of this new PMP criterion in generic letter 89–22, dated Oct. 19, 1989, by requiring that future plants be designed against this new criterion. For existing plants, NRC recommended that the licensees review the material contained in GL 89–22 to determine whether they believe additional action on their part is necessary. However, this previous review was directed toward assessing the adequacy of the design basis, whereas the IPEEE is directed toward severe accidents. Therefore, the staff is requesting that PMP be looked at to assess the effects on plants in terms of onsite flooding and roof ponding to determine whether that would lead to a severe accident. This is consistent with the staff's request that licensees confirm that no plant-unique external events with the potential to initiate severe accidents have been excluded from the IPEEE, as stated in Section 3 of the Generic Letter 88–20, Supplement 4. The general procedure can be found in section D. 2.4 of SRP and section 11.4 of NUREG/CR–2300. The staff believes that this information is readily available per GL 89–22 recommendation.

6. Internal Fires

6.1 The generic letter doesn't state whether the Fire Vulnerability Evaluation (FIVE) methodology and the associated database, being developed by NUMARC and EPRI, are acceptable or not, either for meeting the IPEEE requirements or for satisfying the issues in the Sandia fire risk scoping study.

Are you planning to put that in the final generic letter and guidance document? Since we don't know whether the FIVE methodology is acceptable or not, we can't really make our decision on what method to use and respond in the 60-day timeframe as requested in the generic letter. (Ref. D.16, p. 96)

SR: The development of FIVE has been acknowledged in the generic letter supplement. However, the staff is unable at this time to endorse the FIVE method for the use in IPEEE, because NUMARC and EPRI have not completed its validation and documentation. The staff intends to review the NUMARC/ EPRI methodology and, if it is acceptable, endorse it as an acceptable way to deal with fires under the IPEEE. However, final review will not be completed in time to incorporate a final position on FIVE into the generic letter, therefore, the staff will address the acceptability of the FIVE methodology in a separate document.

With regard to the 60-day initial response time, that has been extended to 180 days after the issuance of the final generic letter, primarily to allow time for completion of the FIVE methodology.

6.2 What are the procedurally directed walk-downs in terms of addressing seismic–fire interaction. Do they pertain to walk-downs for the fire or walk-downs for the seismic IPEEE. (Ref. D.16, p. 130)

SR: The procedurally directed walk-downs associated with internal fires vulnerability evaluation can be

planned as part of the seismic walk-downs that would specifically look for the seismic-induced fire vulnerability issues. The idea is to first identify those areas that could be vulnerable so that they can be brought into focus during the walkdown.

For example, if a plant didn't have its diesel fuel tank strapped down properly one could postulate a large fuel source for fire as a result of a seismic event. Other similar seismic/fire interactions were summarized in Section 7 of NUREG/CR–5088.

6.3 If the utilities had already assessed the safe shutdown capability, are spot checks acceptable for the cable routing verification? (Ref. D.16, p. 130)

SR: Licensees should rely on previous assessments for IPEEE information, provided the information is up-to-date. The licensee does not have to perform any design verification, or retrace cables if that had been done previously.

6.4 Will the fire database be updated and when will it be available? (Ref. D.16, p. 132)

SR: EPRI is updating the fire database through 1988. It is expected to be available to staff sometime in the December–January timeframe.

6.5 How is safety system separation to be assessed in the fire analysis? (Ref. D.16, p. 132)

SR: Separation should be modeled as it exists and the fuel sources as they exist in order to understand, using the codes (propagation analysis), the effects of fire on redundant trains of equipment.

6.6 In treating of transient combustibles for fire, would it be sufficient in the IPEEE fire analysis to state that administrative control for transient combustibles takes care of this or would additional quantitative analysis be needed? (Ref. D.16, p. 148)

SR: Transient combustibles cannot be ignored. If they're procedurally controlled, a fire protection engineer should be involved in doing the examination, so that a determination can be made as to whether procedural control will really limit significant transient combustibles in a given area.

6.7 The requirement stated in Sec. 4.3 Item 5, "provide documentation in each case where the plant-specific data used is less conservative than the approved data base" is a disincentive to use plant specific fire initiation data. The IPEEE results will capture generic vulnerabilities instead of plant specific fire vulnerabilities by preferring generic data (Ref. D.8)

SR: In most cases, plant-specific data are rather limited. The use of generic data, which usually have a much broader base, will provide valuable insights about what could happen at a plant. The awareness of this kind information is very important to the plant operating staff. This is consistent with the purpose of the Commission's Severe Accident Policy, "to understand the most likely severe accident sequences that could occur at its plant."

6.8 Water as a suppression agent may cause potential damage to safe shutdown components. What about CO_2 and Halon? The Sandia study states that there are no data available to quantify damage from these sources. Is a simple statement or identification as potential damage adequate? (Ref. D.16, p. 149)

SR: In the majority of instances seen to date in power plants, water caused the damage. Therefore, the staff believes water damage is the most probable. If there is an area where suppression damage from CO_2 or Halon is likely, however, one should not ignore it.

6.9 A major emphasis of the IPE/IPEEE is to have licensee staff involved to help ensure the most benefit. NUREG-1407 states that should involve engineering judgments of the fire safety experts. Does it mean a fire protection engineer is adequate? (Ref. D.16, p. 166)

SR: The staff would consider a fire protection engineer of a utility to be a fire safety expert.

6.10 The inclusion of GI–57, "Effects of Fire Protection System Actuation on Safety-Related Equipment," was questioned. (Ref. D.1)

SR: The effect of fire suppressants on safety equipment is one of the safety issues identified in the Fire Risk Scoping Study, NUREG/CR–5088. Relevant information on whether the actuated fire protection systems would spray safety-related equipment or not and some protective measures to prevent the safety equipment from being sprayed by fire suppressants, can be collected during the walkdown. The additional effort to collect this information during the IPEEE walkdown should be minimal. We also want to point out that the resolution of this issue does not have to be a part of the IPEEE.

7. Seismic Events

7.1 What is the justification for including Earthquakes in IPEEE? (Ref. D.7)

SR: Based upon the examination of NRC's and industry's plant-specific probabilistic risk assessments (PRAs), the mean core damage frequencies at some plants could be relatively high, in the range of 1E–4 to 1E–6 per year (Table D.3). Many cost-effective improvements that would reduce the potential risk were identified from these PRAs; many were implemented at plants as discussed in Appendix 5 of Supplement 4 to Generic Letter 88–20. This finding is

Table D.3 Summary of PRA Results of Core Damage Frequency (1E–5)

Plant	Total	Total Int'l	Total External	Seismic	Fire	Int'l Flood	Extn'l Floods	High Winds	Lightning
PWR									
Pt. Beach	31.3	13.9	17.4	6.1	3.3	7.7	–	0.4	0.006
Turkey Pt.	23.6	7.1	16.51	.7	7.5	–	4.6	2.4	0.26
St. Lucie	7.44	1.4	6.04	1.3	4.4	–	0.32	–	0.02
ANO 1	17.9	8.8	9.15	7.3	0.58	–	0.72	0.53	0.02
IP2	43.5	6.0	37.5	14.0	19.2	–	–	4.3	–
IP3	15.7	9.0	6.7	0.31	6.3	–	–	0.13	–
Zion	34–40	34.2	0.1–6	<0.1–6					
MS3	15–23	14.7	0.8–8Est	–	–	–	–	–	
Oconee	15–28	7.4	8–21	6.0	1.0	–	0	1.3	–
					10.0 (NRC)	–	2.5 (NRC)	2.3 (NRC)	–
BWR									
Quad City	19.7	9.9	9.8	8.3	1.3	–	0.01	0.01	0.2
Cooper	43.7	28.9	14.8	8.1	1.1	–	5.	0.4	0.2
Limerick	9.2	8.4	0.8	0.5	0.3	–	–	–	–
Shoreham	7.4	5.4	2.0	–	–	2.0 (NRC)	–	–	

IP2/IP3—Indian Point 2/Indian Point 3
MS3—Millstone 3

consistent with the statement in the Commission's Severe Accident Policy that systematic examinations are beneficial in identifying plant-specific vulnerabilities to severe accidents that could be fixed with low-cost improvements. References D.2, D.4, D.6, and D.11 all acknowledged that systematic examinations are valuable tools for gaining insights into the plant operation and identifying cost-effective plant improvements.

Another indication that earthquakes can be important risk contributors can be seen from the earthquake experience on foreign nuclear power plants. For example, (1) On April 22, 1987, Fukushima units 1, 3, and 5 in Japan, tripped as a result of an earthquake with a magnitude of 6.6; and (2) On March 4, 1977, one steam generator at the Kozloduy nuclear power plant in Bulgaria was displaced by 5 inches. This earthquake experience caused a major overhaul in the seismic design of later Russian plants. It also resulted in a major backfit at the Armenia unit 2 in Soviet Russia.

Therefore, based on risk considerations and the potential for identifying cost-effective improvements, the NRC concludes that seismic events should be included in the IPEEE.

7.2 The use of both Seismic Hazard Curves (LLNL and EPRI Seismic Hazard Curves) was questioned (Refs. D.1, 6, 7, 11, & 20)

SR: There is not enough earthquake data at this time to determine precisely the validity of a single curve. In other words, there is no way to conclusively validate or dispute either curve. Therefore, both the LLNL and EPRI seismic hazard curves are recommended for use in the seismic PRAs. This is consistent with the recommendation of the NUREG–1150 peer review group. An acceptable alternative for licensees choosing to perform only one analysis is to use the higher curve. Also see response 7.8.

7.3 Are extensive margin calculations needed for all components? (Ref. D.7)

SR: Extensive calculations of HCLPFs are not needed in order to achieve the NRC's goals for the seismic IPEEE. Refer to response 7.17 for the scope of margins evaluations. Also, see revised Sections 3.2.4.6 and 3.2.5.7 of this report.

7.4 IPEEE Objectives and Methods: ". . . recommend that the objectives of the IPEEE be modified to better delineate IPEEE objectives for each of the accepted methodologies, or other non-probabilistic methods that may be proposed by licensees." (Ref. D.1)

SR: References to probabilities are now eliminated from the objectives. The IPEEE objective is to gain a qualitative understanding of the overall likelihood of core damage and fission product releases. This is different from that of the IPE, where quantitative understanding is emphasized. In Section 1.2 of this report, the staff has acknowledged specifically that "some methods have been developed for evaluating external hazards and identifying vulnerabilities that do not produce estimates of core damage frequency. . . .Thus, objective 3 above would be addressed only indirectly for some methods to be used in the IPEEE" It should be noted that the seismic margin methods were derived from seismic PRAs. The objectives have been reworded to emphasize this point.

Also, see the staff response to items 1.11 and 1.14.

7.5 Scope and Depth of Assessment: "The scope of evaluations requested for seismic IPE is more extensive than needed to satisfy the Severe Accident Policy Statement." (Ref. D.1)

SR: The staff, based on a NUMARC recommendation, has defined three review categories with varying levels of effort. This approach leads to an overall reduction in the effort to carry out the examination. Additional details are discussed in response to comment 7.17.

7.6 Seismically induced floods are mentioned for the first time in draft NUREG–1407, Section 6.3.2 and not in Generic Letter 88–20 Supplement 4. We understand that the scope of review for seismically induced external flooding is limited to a review of external sources of water (e.g., tanks, upstream dams, or other significant structures) and not internal water sources such as piping. This should be clearly stated in Generic Letter 88–20, Supplement 4, in order to avoid possible confusion in future interpretations. (Ref. D.1)

SR: The scope of the seismically induced floods, in addition to the external sources, includes the evaluation of some internal flooding consistent with the discussion in Appendix I, Check Lists and Walkdown Data Sheets, of EPRI NP-6041. Section 6.3.2 will be modified to include reference to EPRI NP-6041. In addition, the generic letter has been modified.

7.7 Draft Generic letter 88–20, Supplement 4, Section 5 lists the three related programs subsumed in the IPEEE: (1) the external event portion of USI A–45, "Shutdown Decay Heat Removal Requirements," (2) GI–131, "Potential Seismic Interaction Involving the Movable In-Core Flux Mapping System Used in Westinghouse Plants," and (3) the "Charleston Earthquake Issue." When the IPEEE is utilized for closure of a subsumed issue, we understand that no special evaluation, documentation, or reporting will be needed beyond the program defined by IPEEE (Ref. D.1).

SR: It is generally true that evaluation and reporting beyond that identified in this report should not be required for subsumed issues except for any additional information that may be needed as a part of the normal staff review of the IPEEE submittal. Note though, that the IPEEE submittal is to address specifically USI A–45 and GI–131. Closure of the IPEEE also means closure for these issues. No specific reporting requirements are identified for the Eastern U.S. Seismicity Issue, formerly identified as the Charleston Earthquake Issue.

7.8 Seismic hazard related comments from Ref. D.1, p 8, Comment 1, are summarized as follows:

a. Use of two hazard curves is not needed. Additional expenses for two analyses not justified.

b. The EPRI methodology has been reviewed and accepted by the USGS. The LLNL results are not realistic.

SR: a. The staff has revised its position regarding the use of hazard curves for a PRA analysis in response to this and other similar comments. The staff still prefers that both LLNL and EPRI hazard curves be used in an analysis as this will serve to highlight uncertainties in the bottom line numbers as well as robustness in the identification of vulnerabilities. However, an option of using one hazard curve is now included provided the higher hazard curve of the two is used. The reasons for using the higher hazard curve are twofold. One, as discussed in more detail in b. below, is that the validity of one curve over the other has not been determined yet. The second is that the use of the higher hazard curve will ensure that all potential seismic sequences are identified.

Comments related to expenses appear to result, to a certain extent, from a misunderstanding of the scope of analyses required to obtain results using two hazard curves. Similar comments were also made at the workshop specifically suggesting that two hazard curves will necessitate two separate plant response and fragility analyses. However, the staff never intended that two separate response or fragility analyses would be needed and therefore specified only one spectral shape. This point is further emphasized in Section 3.1.1.2 of this report by categorically stating: "Since only one spectral shape is used for both hazard analyses,

two separate plant response and fragility analyses are not needed." The additional effort to generate results from an additional hazard curve is relatively trivial, and requires convolution of a hazard curve with the existing plant level and sequence level fragility curves.

b. Licensees east of the Rocky Mountains using a seismic PRA for the IPEEE examination are requested to use the results of two seismic hazard studies. These studies, conducted by the Lawrence Livermore National Laboratory (LLNL) and the Electric Power Research Institute (EPRI), represent a state-of-the-art developmental effort. However, for reasons associated with methodology development, these two studies can produce significantly different seismic hazard curves.

The uncertainty associated with seismic hazard assessment is clearly demonstrated when one compares the vast differences between the mean, median, 15th, and 85th percentile estimates associated with one curve. The differences between the LLNL and EPRI hazard curves further demonstrate the large uncertainty.

In an attempt to resolve the differences between the two curves the staff requested assistance from the National Academy of Sciences. The Academy has criteria it uses to evaluate scientific and technical approaches used in research. It concluded that both studies followed good scientific procedures and practices, and therefore both studies are credible. Therefore, the staff is encouraging licensees to use both LLNL and EPRI seismic hazard curves in the IPEEE evaluation. However, if only one curve is used, it should be the higher one. As stated earlier, the use of both will serve to identify differences, if any, in the delineation of dominant seismic sequences. Although NUREG-1150 studies did not identify any significant differences in ranking and contributions, variations in contributions and rankings could occur when curves have markedly different slope characteristics. Taken together, these pieces of data (contributions and rankings from both hazard estimates) give a complete representation of the seismic event. These data will be extremely useful to the licensee in identifying plant vulnerabilities and deciding if plant modifications are warranted.

The NUMARC claim about the review and acceptance of the EPRI methodology by USGS appears to mischaracterize the extent and intent of the USGS review. Certain observations need to be made regarding the USGS review of the EPRI methodology. First of all, the USGS reviewed the EPRI methodology as a staff contractor and their findings were incorporated in the staff Safety Evaluation Report (SER) addressing this methodology. Second, the review was limited to the methodology and did not include geology, tectonics, ground motion, or site specific results. The staff SER (Richardson, 1988) clearly stated that acceptance was limited to the methodology, and any application to regulatory issues was not part of the approval. Finally, there were a number of caveats in both the USGS and the staff evaluations and neither indicated a blanket acceptance of the EPRI results. The staff conditioned its approval by noting several areas in which problems may arise if certain precautions were not observed based on the USGS review. The staff concluded that ". . . the staff intends to use seismic hazard calculations resulting from the application of the SOG/ EPRI methodology in conjunction with similar results obtained from LLNL Seismic Hazard Characterization Program (SHCP). If significant differences are observed that can not be resolved, the NRC staff will use the two sets of calculations to define the range of seismic hazard to be used in the decision making process. In any case, these uncertainties are such that the specific calculation of seismic hazard, be it that obtained by EPRI or LLNL, should be viewed with some caution. The staff finds that seismic hazard calculations are better used for making relative comparisons than for placing reliance upon the specific numerical estimates."

7.9 The two Eastern U.S. plants that were placed in the 0.5g RLE bin need not be reviewed at that level. Recommend adding a footnote that should read ". . . indicates an Eastern U.S. Site whose RLE is greater than 0.3g unless the licensee can demonstrate on site specific examination that the plant's seismic exposure is similar to, or less than, those plants assigned to 0.3g RLE". (Refs. D.1, 10, and 14)

SR: Since the publication of the draft generic letter and the guidance document, both plants placed in the 0.5g RLE bin have committed to either enhance the existing PRA (Ref. D.2) or to perform a new seismic PRA (Ref. D.19); therefore, they have been removed from this category.

7.10 Reporting of HCLPF values for components, sequences, and the plant for both new and existing PRAs should not be required. (Ref. D.1)

SR: The staff has accepted this comment and the reporting of HCLPF values for licensees using PRA methods is now optional. The staff has also indicated that this information is readily available from PRAs, and

the staff intends to use HCLPF or margin related insights in the evaluation of the IPEEE submittals. It should be further noted that it is a current practice to include HCLPF information in PRA submittals.

7.11 Draft Generic Letter 88–20, Supplement 4, Appendix 4, Section 4.2.2, Item 6, for SMA method, calls for calculation of HCLPF values "with or without nonseismic failures and human actions." This item should be clarified to state that it does not apply to the EPRI SMA methodology. We understand that was the intent because, in the EPRI SMA method, success paths are chosen avoiding unreliable equipment and unrealistic human recovery actions. (Ref. D.1)

SR: The staff agrees and will clarify the noted item. Also, see response to item 7.6 of Section D.1 of this Appendix.

7.12 With regard to the containment performance evaluation (Ref. D.1):

a. It should be clarified that only systems required to prevent early containment failure need to be assessed.

b. For reduced-scope plants, we recommend no containment performance evaluation.

SR: a. The staff has now included this clarification (see the staff response to item 7.7 of Section D.1 of this Appendix).

b. The staff is still recommending retention of the walkdown of containment systems necessary to prevent early failures because the walkdown will identify anchorage and spatial interaction problems that can occur. Such a review of containment is consistent with the defense-in-depth philosophy adopted in other parts of this program.

7.13 The draft generic letter states that the Charleston earthquake issue is subsumed in the IPEEE and that completion of the IPEEE will constitute a resolution of the Charleston earthquake issue. We believe the Charleston earthquake issue should be closed based on the information contained in EPRI Report, NP–6395–D. (Ref. D.1, & Ref. D.5, Attachment 1)

SR: The issue of the 1886 Charleston earthquake has been resolved. The issue of eight outlier plants identified thru the Eastern U.S. Seismicity program has been subsumed in the IPEEE and no specific reporting is required to close this issue. The staff will review the IPEEE results for the affected plants.

7.14 Median estimates of seismic hazard curves should be used rather than mean values. Mean values are unduly influenced by outlier experts and thus are unstable. Median values are less affected by the extreme estimates and thus provide more stability. (Ref. D.1)

SR: The staff is recommending the use of mean hazard curves for the following reasons: (1) The use of mean hazard curves and mean fragility curves will lead to approximate mean level frequencies for core damage. No statistical meaning can be attached to a point estimate obtained through use of median hazard curves. (2) The instability and uncertainties are better presented by displaying mean results from two hazard studies as recommended by the staff. The use of median curves is tantamount to ignoring uncertainties and some expert opinions without an adequate technical basis.

7.15 Scope of relay chatter evaluation. (Ref. D.1)

SR: See the staff response to item 7.17 of this section, response to Attachment 1 of Ref. D.1.

7.16 If the intent for the NRC SMA method is to require the development of level 1 and 2 functional sequences from event trees, the cost of SMA using the NRC method would be substantially increased. (Ref. D.1)

SR: The staff does not require the development of functional event trees beyond that contained in NUREG/CR-4334, NUREG/CR-4482, and NUREG/CR-5076, to address containment performance issues when the margin approach is used. (Note that for the PRA approach the tie-in between Level 1 and Level 2 is quite clear, and this should not be an issue). As stated in this report, the licensee should develop its own containment performance plans based on the IPE results. What is required is to examine containment functions (regardless of the plant damage states that may be indicated by Level 1 margin sequences or success paths) required to prevent early failures and report HCLPFs for these functions and components if below the RLE. Obviously, a licensee has an option to develop Level 2 trees at its discretion. Discussions of various ways to extend Level 1 margin analysis to Level 2 are contained in Budnitz 1991a and 1991b for both NRC and EPRI methods. (This is suggested for general guidance, no specific requirements based on these references are implied.) A success path oriented approach is also discussed in Ref. D.1 and Reed, et al., 1990. In summary, the cost of the NRC SMA method need not be greater than the EPRI SMA method.

7.17 Summary of NUMARC recommendations for the implementation of the seismic aspects of Generic Letter 88-20, Supplement 4. (Ref. D.1, Attachment 1)

SR: The staff responses to these summary recommendations are provided section by section in accordance with Attachment 1 of Ref. D.1. Many of these recommendations have been discussed earlier. Detailed discussion is included only when the staff does not agree totally with a recommendation.

Selection of Full-, Focused- and Reduced-Scope Plants

Reduced-scope, 0.3g RLE Review (Full-scope and Focused-scope SMA):

The staff has accepted the NUMARC suggestion of creating full- and focused-scope categories in the 0.3g bin. The 0.3g bin is subdivided into the full and focused scope based on the NUMARC suggested approach of using both hazard and seismic design basis as parameters. Additional consideration was also given to the identification of outlier plants resulting from resolution of the Charleston earthquake issue (see Appendix A of this report for more discussion).

.1 *0.5g RLE Review:* Provide opportunity to two Eastern U.S. plants in the 0.5g bin to submit site-specific justification for a binning change from 0.5g to 0.3g, similar to consideration given to Western US plants.

SR: See response to item 7.9 of this section.

.2 *Multiple Units at a Site:* Lessons learned in evaluating the first unit may be used in examining the other unit(s); in particular, any areas of concern that may be identified during the evaluation of the first unit would be examined in the other unit(s). Otherwise, the scope of review for the other unit(s) can be reduced accordingly.

SR: The staff agrees that results and findings from the first unit should be used to help in the evaluation of other units, provided appropriate similarities exist. However, such judgements can be made only on a case-by-case basis. This report has been revised to include a statement to this effect. In any event, walkdowns of all units will have to take place to ensure similarities. It is very likely that the greatest reduction would be achievable in analytical effort.

.3 *Scope of Deterministic Seismic Review:* Identification of Success Path Elements. For all three types of review (full-, focused-, and reduced-scope), procedures for identifying structures and equipment to be reviewed are the same and are based on the recommendations in EPRI Report NP–6041.

SR: The staff agrees and has referenced EPRI NP–6041 as the primary document for the EPRI success path methodology. However, due consideration should be given to supplemental comments made in Sections 3.2.5.1 and 3.2.5.8 of this report regarding the selection of success paths.

.4 *Containment Review:* The full- and focused-scope SMA reviews should be limited to evaluation of only those functions that are necessary to prevent early containment failure.

SR: See the staff response to item 7.12 of this section.

.5 *Relay Evaluation:* The following table outlines the NUMARC recommended position on relay chatter evaluation. This table is based on presentation made to the staff by NUMARC on November 29, 1990.

Table 7.17.1 NUMARC Recommendations for Relay Chatter Review

Review Type	Plant Type	Recommended Review
Full-scope	A–46	Evaluate A–46 per A–46. For relays within IPEEE (not in A–46), perform a bad actors review.
	Non A–46	Perform a bad actors review for all relays within IPEEE.
Focused-scope	A–46	Evaluate A–46 relays per A–46 (SSE). If bad actors are found, expand scope to include IPEEE relays.
	Non A–46	Perform a bad actors review for all relays within IPEEE.
Reduced-scope	A–46	Perform A–46 review. No additional review for IPEEE relays.
	Non A–46	No relay evaluation.

SR: The staff recommended relay chatter evaluation is outlined in the following table.

Table 7.17.2 NRC Recommendations for Relay Chatter Review

Review Type	Plant Type	Recommended Review
Full-scope	A–46	Follow A–46 procedures for A–46 review. Expand scope to include IPEEE systems using appropriate margin or A–46 procedure. Review at assigned RLE.
	Non A–46	Review all IPEEE systems using appropriate procedures at RLE.
Focused-scope	A–46	Same as NUMARC recommendation.
	Non A–46	Same as NUMARC recommendation.
Reduced-scope	A–46	Same as NUMARC recommendation.
	Non A–46	Same as NUMARC recommendation.

Comparison between Tables 7.17.1 and 7.17.2 indicates that the staff is in agreement with NUMARC on the focused- and reduced-scope categories encompassing the majority of plants. Reasons for differences in the full-scope review are discussed in Section 3.2.1 of this report. It should be noted that, for plants performing PRAs, the scope of the relay review is also defined by the above table.

.6 *Soil Failure Investigation:* For plants in the focused-scope SMA category, a review based on the design and construction record is considered adequate. A review of soil failure should not be required for plants in the reduced-scope bin.

SR: The staff has adopted both recommendations. For the focused-scope category, the use of design and construction records is considered adequate provided appropriate data is available. A detailed analysis will be performed at the licensee's discretion if soil failure is found to be significant. For the reduced-scope bin, no soil evaluation is required. However, it should be noted that the need and the effort required to evaluate soil failure is site specific and should be determined on a case-by-case basis.

.7 *Screening Criteria:* Tables 2–3 and 2–4 of EPRI NP–6041 can be used. The A–46 screening guidance given in the Generic Implementation Procedure (GIP) may also be used.

SR: The staff has adopted this recommendation. It should be noted that all caveats given in the margin methodology as well as limitations on the use of GERs should be observed, and the IPEEE review is to be performed at the assigned RLE. Spatial interaction issues, such as flooding discussed in EPRI NP–6041, must be addressed.

.8 *Evaluation of Outliers:* For both full- and focused-scope SMA reviews, HCLPFs should be determined for elements not screened out during a walkdown. For focused-scope reviews, it is recommended that judgement be used to rank the capacities of the outliers from the lowest to the highest. HCLPF capacities should be calculated as necessary for some components, other components should be assigned conservative HCLPFs. For reduced-scope plants, outliers should be evaluated according to the plant FSAR.

SR: The staff has adopted these recommendations. See Sections 3.2.4.6 and 3.2.5.5 of this report.

.9 *Seismic Input:* For full- and focused-scope SMA reviews, use NUREG–0098 median spectra anchored to the RLE for the plant. For reduced-scope reviews, use spectra developed for the SSE ground response spectrum.

SR: The NUMARC recommendation of the use of NUREG–0098 is consistent with the staff recommendation. The suggested recommendation for the reduced-scope review is accepted with a caveat that any difference between FSAR and new response spectra should be highlighted and discussed.

.10 *Review Documentation:* The documentation of the IPEEE for the full- and focused-scope SMA review should follow the guidance outlined in EPRI Report NP–6041. The report for the reduced-scope review should be concise.

SR: The staff expects that information outlined in Appendix C will be included in the IPEEE submittals.

Integration of IPEEE and A–46 Reviews:

.11 It is recommended that IPEEE and USI A–46 reviews be conducted concurrently and that the review tasks be combined whenever possible.

SR: The staff welcomes the NUMARC emphasis on integrating these two major seismic efforts; this recommendation is consistent with the staff philosophy discussed in Section 6 of this report.

Scope of Seismic Review Using SPRA Approach:

.12 *Use of Seismic Hazard Results:*

a. Hazard results presented in EPRI Report NP–6395–D can be used in performing the SPRA.

b. NRC should allow licensees an opportunity to perform site specific studies in order to develop new, more realistic seismic hazard data.

SR: a. See the staff response to item 7.8 in this section.

b. Licensees always have the option to conduct additional studies they deem necessary and present them to the staff for review. However, the new hazard should not be used in lieu of the LLNL hazard, but it can be used to provide additional insight into uncertainties.

.13 *Fragility Calculations:* Mean fragility curves are adequate for fragility calculations.

SR: This is identical to the staff position.

.14 *Relay Chatter:* Consideration of relays in a seismic PRA should be limited to relays with low seismic ruggedness.

SR: The staff preference is that the relay chatter review scope be defined by the plant categorization used in the margin review. That is, for a plant identified in the full-scope category, if the licensee chooses to conduct a seismic PRA, the relay review is to be done as outlined for that category. Relay fragilities and recovery actions should be modeled in the PRA as appropriate.

.15 *HCLPF Calculations:* HCLPF calculations should not be required for a SPRA.

SR: The staff has made this an optional recommendation. See response to item 7.10 of this section.

7.18 The New Hampshire Yankee and Boston Edison requested that the NRC recognize their use of PRA for performing the seismic portion of the IPEEE, in both NUREG–1407 and GL 88–20, Supplement 4, prior to final issuance. (Refs. D.2 & 19)

SR: The staff has modified GL 88–20, Supplement 4, and this report to acknowledge the licensee's commitment to use PRA.

7.19 An alternate binning approach, using both hazard and seismic design basis considerations, should be considered. (Ref. D.4)

SR: From several suggestions regarding the binning process, the staff has accepted the binning process recommended by the NUMARC to further subdivide the 0.3g bin plants into the full- and focused-scopes. Many of the individual utilities have endorsed the NUMARC comments.

7.20 Utilities should be given the option of using either LLNL or EPRI hazard results. (Ref. D.5, Attachment 1)

SR: See the staff response to item 7.8 of this section.

7.21 The IPEEE should not be required for closure of Charleston for every plant. (Ref. D.5)

SR: See the staff response to item 7.13 of this section.

7.22 The NRC should consider modifying the bin categories based on the design hazard concept proposed jointly by NUMARC/EPRI. An alternate binning scheme is suggested. (Ref. D.5)

SR: The staff has considered the NUMARC/EPRI approach in further subdividing the 0.3g bin. The staff binning approach is more consistent with the NUMARC suggested approach in Ref. D.1.

7.23 Clarify the extent of peer review for the seismic IPEEE. (Ref. D.5)

SR: The staff intent is now clarified in Sections 3 and 7 of this report and the generic letter, the extent of the peer review should be consistent with the IPE guidance as provided in NUREG-1335.

7.24 The use of two hazard curves is illogical. Allow the use of EPRI hazard data. (Ref. D.5)

SR: See the staff response to item 7.8 of this section.

7.25 Recommend deleting the containment walkdown for reduced-scope studies. (Ref. D.5)

SR: See the staff response to item 7.12 of this section.

7.26 In Section 3.2.6 there is some confusion. If you utilize the IPE to identify "success paths," then you would not identify sequences and seismic failure modes that are significantly different from those found in the IPE internal event evaluation. (Ref. D.5)

SR: Even if the IPE is used to identify success paths, failure modes such as passive failures, structural failures, and spatial interaction failures are generally not considered in an internal event IPE. Additionally, the "common cause" effect created by a seismic event is unique in that the entire plant is subject to the ground motion causing combinations of failures that may not be manifested in an internal event IPE. Thus different failure modes and combi-

nations of failures can induce sequences that are different from those found in the IPE internal event evaluation.

7.27 The Charleston issue should be closed for a majority of the Eastern U.S. plants; for the outliers, the issue can be subsumed through the IPEEE.

SR: The staff agrees. See response to item 7.13 of this section.

7.28 Does Section 6.3.2 imply that seismic event success paths must also be simultaneously protected from postulated fire/floods? The sentence "The effects of seismically induced external flooding and internal flooding on plant safety should be included" is not clear. (Ref. D.5)

SR: With regard to floods, see the comment and the staff response to item 7.6 of this section.

With regard to fire, see the staff response to item 6.2 of this section.

7.29 The sentence "However, the licensee should assess the significance of HCLPF values lower than RLE and take any necessary actions and make other improvements that are deemed appropriate by the licensee." is too arbitrary. More specific guidance is necessary. (Ref. D.5)

SR: The judgements about the significance of findings can be made only when findings are available; therefore, more specific guidance is difficult to give at this time, and such attempts may create confusion. However, the intent of the statement is to limit the scope of evaluation for which significance needs to be assessed.

7.30 The staff binning process only recognizes hazard and not design. (Ref. D.5)

SR: See the staff response to item 7.22 of this section.

7.31 Provide additional guidance about some twelve plant sites east of the Rocky Mountains whose main Category I structures are located on rock, and also have some Category I structures or components located on shallow or intermediate depths of soil. (Ref. D.5)

SR: The RLE assignment has been made considering soil conditions where the main plant structures are located, namely, at rock level for the above cases. As noted in this report, significant amplification may occur through the soil layers above the rock, and, hence, plant structures founded on soil may experience much higher motion than the rock-founded structures. In such cases, the use of screening tables based on the RLE assignment may not be appropriate for the soil-founded structures and components. The licensee should investigate this soil amplification phenomenon using any suitable means (e.g., analytical studies, comparisons with other appropriate studies) to determine how to evaluate the soil-founded structures and components.

7.32 Suggests an alternate approach related to seismic binning. (Refs. D.5 and 10)

SR: See the staff responses to items 7.17 and 7.19 of this section.

7.33 The tie-in of seismic margin to a Level 2 PRA is not defined. Also, clarify "all HCLPFs related to...containment performance" (Ref. D.6, Enclosure)

SR: See the staff response to item 7.16.

7.34 Does resolution of USI A–45 have scope implications for seismic margin options. (Ref. D.6, Enclosure)

SR: Yes, as noted in Section 6.3.3.1, functions and systems for addressing USI A–45 should be same as those identified in the internal event IPE. Otherwise, some of these functions and systems may not necessarily be included in a margin evaluation. A specific reporting provision is also called out in item 7 of Section C.2.2, Appendix C, of this report for this USI.

7.35 On page 24, Section 4.2.2, Item 2 of the draft generic letter, replace the term "findings" with "results." (Ref. D.8)

SR: The staff accepts this comment; appropriate sections are revised to reflect this change.

7.36 A peer review implying the use of external "experts in the professional field" for a review of the methodology chosen and its application is not necessary. in-house review is more appropriate. (Ref. D.8)

SR: The staff has now clarified peer review discussions in the generic letter and this report to be consistent with the internal event IPE guidance (NUREG-1335), which emphasizes in-house review. However, the staff has also recognized that a licensee may not have in-house expertise in all areas of the external events and an in-house team can be supplemented by outside experts.

Furthermore, it should be recognized that substantial judgement is involved in applications of PRA as

well as margin methodologies as demonstrated in trial applications at Maine Yankee (NUREG/CR–4826) and Hatch (Davis, 1990). This is particularly important now that the "focused-scope" category has been introduced requiring more use of judgement. The composition of the in-house team should therefore strike a balance so that sufficient expertise is available to ensure that the methodology is properly implemented while utilizing in-house staff as much as possible.

7.37 Based on lessons learned from PRA and Margin evaluations a simplified walkdown procedure for a majority of plants should be developed. (Ref. D.9)

SR: The staff has revised the scope of the seismic examination in line with NUMARC's suggestions with some exceptions.

7.38 The scope and objective of the walkdown are not sufficiently described. The "40 person months per unit" required for a walkdown is excessive. (Ref. D.18)

SR: No changes to Appendix 1 of the Generic Letter are required. This report was revised to read as follows:

... perform a walkdown consistent with the intent of the guidelines described in Sections 5 and 8, and Appendices D and I of the EPRI Seismic ...

The 40 person month estimate indicated in the comment is not consistent with past experience at several plants. For example, EPRI NP–6359 (Seismic Margin Assessment of the Catawba Nuclear Station) states that the total technical manpower expended was 39.7 man months: approximately 16 man months associated with walkdown preparation and the walkdown, 21.5 man months for evaluating unscreened components, and 2.0 man months for reporting. Southern Company Services, Inc. (Hatch Plant) estimates that they expended 8 staff months for walkdown planning and the walkdown.

The level of effort and resource allocation are justified based on approximately 20 seismic PRAs and 3 seismic margins evaluations. Both evaluation methods have demonstrated that thorough walkdowns are one of the most important tools for identifying seismic weak links.

The importance of a detailed walkdown is also supported by NUMARC. In fact, when NUMARC proposed the Reduced Scope Program for sites in low seismic areas, a plant walkdown identical to the Full Scope evaluation was recommended.

7.39 An analysis using both LLNL and EPRI hazard curves is unnecessary. There is inconsistency between the generic letter and NUREG regarding use of both curves in an existing PRA. (Ref. D.18)

SR: See the staff response to item 7.8 of this section. The generic letter and NUREG have been revised to be consistent.

7.40 To avoid a different interpretation, define HCLPF for the sequences and plant for the PRA and margins methodologies. (Ref. D.18)

SR: The term HCLPF in the context of the margin methods is clearly defined in both NUREG/CR-4334 and EPRI NP-6041. Examples of how plant level and sequence level HCLPFs are determined can be seen in NUREG/CR-4826, the margin evaluation for the Maine Yankee plant. The mathematical definition of HCLPF for both the PRA and margin methods is the same. Examples of determining sequence and plant level HCLPFs are also described in NUREG/CR-4334. Section 3.1.1.3 of this report gives further guidance on how to determine component, sequence, and plant level fragilities when only mean fragility curves are used.

7.41 Soil liquefaction computations to the level of detail recommended in EPRI NP-6041 are not necessary to obtain a qualitative understanding of the overall probability of core damage and radioactive material release.

What is appropriate and reasonable is an assessment as to whether the site is susceptible to liquefaction behavior. (Ref. D.18)

SR: See the staff response to item 7.17, Soil Failure Investigation. The staff has accepted the NUMARC recommendations in this area.

7.42 Put Farley in the Reduced Scope Program bin. (Ref. D.10)

SR: Farley is now assigned to the Focused Scope bin. Plant binning was accomplished by comparing nine separate pieces of information related to seismic hazard groupings and engineering judgement.

7.43 Required use of LLNL and EPRI hazard curves adds significant expense for no significant benefit. The four purposes of the IPEEE can be met using EPRI hazard curves. The EPRI method has been reviewed and accepted by USGS and NRC; the LLNL method has not. (Ref. D.11)

SR: See the staff response to Item 7.8 of this section.

7.44 Reporting of "functional sequences" may not be possible if only systemic sequences are generated for a PRA or the EPRI success path approach is used for a SMA. (Ref. D.11)

SR: True. For those cases, the reporting criteria are described in Appendix 3 of Supplement 4 to GL 88–20.

7.45 Reporting HCLPFs with and without non-seismic failures and human actions does not contribute to the four stated purposes of the IPEEE. (Ref. D.11)

SR: The fourth purpose of the IPEEE is to reduce the overall likelihood of core damage and radioactive material release by modifying hardware and procedures. Cost-effective decisions can be made only if both seismic and non-seismic failures are included in the licensee's decision making process. If the random failure probability of a diesel generator to start is high, no seismic fix to the plant is likely to significantly reduce the frequency of the sequence.

7.46 Using the SMA it is difficult to perform a nonseismic failure and human action evaluation. (Ref. D.11)

SR: Guidance is noted in Sections 3.2.4.7 and 3.2.5.8 on how to address the above evaluations for various methods. Such evaluations have been made and reported in the trial plant applications. Additional guidance on this issue is available in (Budnitz 1987 and 1990). The intent is to help the licensee make the right decisions regarding plant modifications. See response to item 7.46 above.

7.47 Use spectral shapes consistent with the LLNL and EPRI hazard studies. Does this mean uniform hazard spectra? Is this consistent with the NUREG/CR-0098 spectral shape? (Ref. D.11)

SR: This report recommends the use of the median spectral shape for a 10,000 year return period provided in NUREG/CR-5250 or a site specific spectral shape based on a suite of appropriate records for performing PRAs. This is different from the NUREG/CR-0098 spectral shape recommended for evaluations associated with the SMA methodologies. The reasons for this difference is: PRA takes into account the full range of the hazard requiring use of a realistic description of ground motion as much as possible whereas margin evaluations are only conducted at one earthquake level, and the screening tables used in the margin methods are developed from earthquake experience data more compatible with the ground motion represented by the NUREG/CR-0098 spectral shape.

7.48 Allow justification other than expensive sensitivity studies for use of a cutoff other than 1.5g. (Ref. D.11)

SR: Such sensitivity studies are routinely performed to ensure that an adequate range of integration has been defined. These studies are not expensive. In any event, licensees have the option to propose alternative methods. These will be reviewed by the staff on a case-by-case basis.

7.49 A mean component fragility curve is defined by the median capacity, A, and composite uncertainty ... Is this a correct statement? (Ref. D.11)

SR: Yes, the statement is correct.

7.50 Can SQUG GIP walkdown guidelines be used in lieu of EPRI NP–6041? (Ref. D.11)

SR: See the staff response to item 7.17.7, Screening Criteria, of this section.

7.51 Reporting HCLPFs with and without nonseismic failures and human actions does not contribute to the four stated purposes of the IPEEE.

SR: See the staff response to items 7.10 and 7.45 of this section.

7.52 The licensee should have the option of using "heavy duty" experts in lieu of a "peer review" in their SMA. (Ref. D.11)

SR: See the staff response to item 7.36 of this section.

7.53 The Charleston Earthquake Issue does not need to be subsumed into the IPEEE. (Ref. D.11)

SR: See the staff response to item 7.13 of this section.

7.54 Include more Midwest plants in the Reduced-Scope Program bin. (Ref. D.11)

SR: See the staff response to item 7.42 of this section.

7.55 No basis is provided for future plant modifications to maintain the plant margin identified from the IPEEE. (Ref. D.14)

SR: See the staff response in item 8 of Section D.1 of this Appendix.

7.56 There is no specific provision in the Generic Letter to allow a licensee to make their own detailed evaluation to determine which review level earthquake bin they should be assigned. (Ref. D.14)

SR: The assignment of the review level earthquake was based on both state-of-the-art LLNL and EPRI studies. However, the licensee always has an option

to propose an alternative position and submit information to justify it.

7.57 Assignment of the RLE should be allowed to be based upon complete site-specific evaluation of the geological and seismological data for the site. (Ref. D.14, Attachment A)

SR: See the staff response to 7.56 above.

7.58 The cost for Pilgrim will be higher because the staff has characterized the Pilgrim site as a high hazard site. (Ref. D.14, Attachment A)

SR: The cost estimates provided in the draft generic letter are generic. Clearly, some licensees may spend more because more detailed analyses are needed. For instance, containment and containment system performance evaluations for Mark I and Ice Condenser containments will be somewhat more expensive since generic capacity data on the systems are lacking. Licensees have the option of performing a seismic PRA instead of the margins method. Staff consultants have indicated that licensees assigned to the 0.5g bin should seriously consider the PRA option as a means of controlling expenses.

Also, see the staff response to item 7.56 of this section.

7.59 Coordination With Other External Event Programs. Subsumption of the Charleston Earthquake Issue creates a "de facto" provision for IPEEE implementation by pre-supposing that the licensee will be performing a PRA or accepts the NRC's seismic hazard estimates in determining the RLE.

SR: See the staff response to item 7.13.

7.60 The draft Generic Letter should address the future requirements for maintenance of the seismic margin identified by the IPEEE. (Ref. D.14, Attachment A)

SR: See the staff response to item 8 of Section D.1.

7.61 SMA and PRA relay chatter enhancement has not been shown to be cost effective, particularly in light of the A–46 resolution. WE endorses the focused SMA approach for all plants performing a SMA or PRA. (Ref. D.11)

SR: The relay chatter evaluation has been changed; see response to NUMARC's comments in item 7.17 of this section.

7.62 It is not cost beneficial to include long-term cooling and pressure suppression systems in the containment performance evaluation. (Ref. D.11)

SR: The staff has revised the scope of the containment review to focus on the early failure modes. See the staff response to item 7.17.4 in this section.

D.3 References

Budnitz, R., et al., "Extending a HCLPF-Based Seismic Margin Review to Analyze the Potential for Large Radiological Releases and the Importance of Human Factors and Non-Seismic Failures, Draft, March 1987.

——, "Enhancing the NRC and EPRI Seismic Margin Review Methodologies to Analyze the importance of Non-Seismic Failures, Human Errors, Opportunities for Recovery, and Large Radiological Releases, Draft 2, September 1990.

Chilk, S. (NRC), Memorandum to J. Taylor (NRC), and W. Parler (NRC), dated July 17, 1990, Subject: SECY-90-192—Individual Plant Examination for Severe Accident Vulnerabilities due to External Events (IPEEE).

Code of Federal Regulations, Title 10 "Energy" (10 CFR), U.S. Government Printing Office, Washington, D.C., revised periodically.

Davis, P., "A Peer Review of Two Seismic Margin Assessments as Applied to the Hatch Nuclear Power Plant", Proceedings of Third Symposium on Current Issues Related to Nuclear Power Plant Structures, Equipment and Piping, Orlando, Florida, December 1990.

Electric Power Research Institute, EPRI NP-6041, "A Methodology for Assessment of Nuclear Power Plant Seismic Margin," October 1988.

——, EPRI NP-6359, "Seismic Margin Assessment of the Catawba Nuclear Station," Vols. 1 and 2, April 1989.

——, EPRI NP-6395-D, "Probabilistic Seismic Hazard Evaluation at Nuclear Plant Sites in the Central and Eastern United States: Resolution of the Charleston Issue," April 1989.

Moore, D., et al., "Results of the Seismic Margin Assessment of Hatch Nuclear Power Plant," Proceedings of Third Symposium on Current Issues Related to Nuclear Power Plant Structures, Equipment and Piping, Orlando, Florida, December 1990.

Orvis, D., et al, "Seismic Margin Review of Plant Hatch Unit 1: System Analysis," LLNL Report No. ULRL-CR-104834, August 1990.

Rasin, W. (NUMARC), letter to W. Minners (NRC), Subject: Final Industry Comments on Draft Generic Letter 88-20, Supplement 4, "Individual Plant Examination of External Events (IPEEE) for Severe Accident Vulnerabilities," and Draft NUREG-1407, "Procedural and Submittal Guidance for the IPEEE," October 10, 1990.

Reed, J., et al. "Recommended Seismic IPE Resolution Procedure," Proceedings of Third Symposium on Current Issues Related to Nuclear Power Plant Structures, Equipment and Piping, Orlando, Florida, December 1990.

USNRC, Generic Letter 88-20, "Individual Plant Examination for Severe Accident Vulnerabilities—10 CFR 50.54(f)," November 23, 1988.

——, Generic Letter 88-20, Supplement No. 1. "Initiation of the Individual Plant Examination for Severe Accident Vulnerabilities—10 CFR 50.54(f)," August 29, 1989

——, Generic Letter 88-20, Supplement No. 4, "Individual Plant Examination of External Events (IPEEE) for Severe Accident Vulnerabilities—10 CFR 50.54(f)," draft for comment, July 23, 1990.

——, Generic Letter 88-20, Supplement No. 4, "Individual Plant Examination of External Events (IPEEE) for Severe Accident Vulnerabilities—10 CFR 50.54(f)," final, June, 1991.

——, Generic Letter 89-22, "Resolution of Generic Safety Issue No. 103, 'Design for Probable Maximum Precipitation,'" October 19, 1989.

——, NUREG-1032, "Evaluation of Station Blackout Accidents at Nuclear Power Plants," June 1988

——, NUREG-1150, "Severe Accident Risks: An Assessment for Five U.S. Nuclear Power Plants," Vols. 1 and 2, June 1989.

——, NUREG-1335, "Individual Plant Examination: Submittal Guidance," final report, August 1989.

——, NUREG-75/087, "Standard Review Plan for the Review of Safety Analysis Report for Nuclear Power Plants," LWR edition, December 1975.

——, NUREG/CR-0098, "Development of Criteria for Seismic Review of Selected Nuclear Power Plants," May 1978.

——, NUREG/CR-2300, "PRA Procedures Guide," January 1983.

——, NUREG/CR-2815, "Probabilistic Safety Analysis Procedures Guide," Vols. 1 and 2, August 1985.

——, NUREG/CR-4334, "An Approach to the Quantification of Seismic Margins in Nuclear Power Plants," August 1985.

——, NUREG/CR–4482, "Recommendations to the Nuclear Regulatory Commission on Trial Guidelines for Seismic Margin Reviews of Nuclear Power Plants," March 1986.

——, NUREG/CR–4826, "Seismic Margin Review of the Maine Yankee Atomic Power Station," Vols. 1–3, LLNL March 1987.

——, NUREG/CR–5076, "An Approach to the Quantification of Seismic Margins in Nuclear Power Plants: The Importance of BWR Plant Systems and Functions to Seismic Margins," May 1988.

——, NUREG/CR–5088, "Fire Risk Scoping Study," January 1989.

——, NUREG/CR–5250, "Seismic Hazard Characterization of 69 Nuclear Power Plant Sites East of the Rocky Mountains," Vols. 1–8, January 1989.

——, "Policy Statement on Severe Reactor Accidents". Federal Register, Vol. 50, p. 32138, August 8, 1985.

——, SECY 88–147, "Integration Plans for Closure of Severe Accident Issues, May 25, 1988.

D.4 References for Sources of Comments

D.1 Letter from W. Rasin (NUMARC) to W. Minners (NRC), dated October 10, 1990.

D.2 Letter from T. Feigenbaum (New Hampshire Yankee) to Document Control Desk (NRC), dated October 3, 1990.

D.3 Letter from W. Orser (Detroit Edison) to W. Minners (NRC), dated October 8, 1990.

D.4 Letter from R. Wheaton (R. D. Wheaton Asso.) to J. T. Chen (NRC), dated October 6, 1990.

D.5 Letter from M. Lyster (Centerior Energy) to Document Control Desk (NRC), dated October 8, 1990.

D.6 Letter from J. Garrick (PLG Inc.) to J. T. Chen (NRC), dated October 9, 1990.

D.7 Letter from P. Smith (The Readiness Operation) to J. T. Chen (NRC), dated October 4, 1990.

D.8 Letter from G. Muench (Entergy Operation, Inc.) to S. Chilk (NRC), dated October 5, 1990.

D.9 Letter from J. Skolds (South Carolina Electric & Gas Co.) to J. T. Chen (NRC), dated October 10, 1990.

D.10 Letter from W. Hairston, III (Alabama Power Co.) to W. Minners (NRC), dated October 12, 1990.

D.11 Letter from C. Fay (Wisconsin Electric Power Co.) to E. Beckjord (NRC), dated October 5, 1990.

D.12 Letter from G. Sorensen (Washington Public Power Supply System) to Document Control Desk (NRC), dated October 5, 1990.

D.13 Letter from N. Reynolds (Winston & Strawn) to E. Beckjord (NRC), dated October 9, 1990.

D.14 Letter from R. Bird (Boston Edison) to Document Control Desk (NRC), dated October 12, 1990.

D.15 Letter from D. Shelton (Centerior Energy) to Document Control Desk (NRC), dated October 16, 1990.

D.16 USNRC Transcript on IPEEE Workshop, Pittsburgh, Pennsylvania, September 11–13, 1990.

D.17 Letter from G. Goering of Northern States Power Company to F. Miraglia of NRC, dated June 10, 1985.

D.18 Letter from H. Tucker (Duke Power Co.) to E. Beckjord (NRC), dated August 30, 1990.

D.19 Letter from G. Davis (Boston Edison) to Document Control Desk (NRC), dated March 1, 1991.

ATTACHMENT TO APPENDIX D

VALUE/IMPACT ANALYSIS
FOR THE IMPLEMENTATION
OF INDIVIDUAL PLANT EXAMINATION OF EXTERNAL EVENTS

1. Introduction

The primary objective of this attachment is to provide a value/impact analysis to support the issuance of a supplement to Generic Letter 88–20 (Ref. 1) requesting an Individual Plant Examination of External Events (IPEEE) from all licensees holding operating licensees for nuclear power plants. Implementation of the IPEEE program is consistent with the Commission's Severe Accident Policy (50 FR 32138) dated August 8, 1985 (Ref. 2). The implementation of the IPEEE program will provide the utilities and the NRC staff with a better understanding of the actual state of the plant and its capability to cope with severe accidents. The IPEEE program may reveal external event vulnerabilities that could be reduced by procedure changes or hardware modifications to upgrade the frontline and support safety systems.

In general, in performing a value/impact analysis (Ref. 3) the staff would (1) identify potential external event vulnerabilities to severe accidents in operating light water reactor (LWR) power plants, (2) identify modifications that could reduce plant risk from these vulnerabilities, (3) determine the safety benefit of these modifications, and (4) assess the net cost of the modifications. However, in this study, we do not know what the utilities will find from their IPEEE programs. Also, we do not know what fixes the utilities will propose. Therefore, for this study we have used data from published probabilistic risk assessments (PRAs) to identify potential vulnerabilities that could be identified in an IPEEE and compared the benefit of fixing such vulnerabilities to the cost of doing the IPEEE as well as the cost of the fixes themselves.

2. PRA Findings

Although the Commission has concluded that existing plants pose no undue risk to the public, the Commission emphasized that systematic examinations of existing plants are needed to confirm the absence of any plant unique vulnerabilities to severe accidents. This conclusion was based on the fact that previous plant-specific (PRAs) have typically revealed valuable insights on plant specific vulnerabilities to severe accidents.

Table 1 summarizes previous PRA results in terms of core damage frequencies (CDFs) due to internal and external events from 13 PRAs (Refs. 4 & 5) that are available to NRC. These results indicate that the mean value of the CDF, for these plants is in the range of 1E–4 to 4.4E–4 per reactor-year, with the core damage frequencies due to external events being in the range from 6E–5 to 3.8E–4 per reactor year. More recently, NUREG–1150 analyses for Surry (NUREG/CR–4551, Volume 3) and Peach Bottom (NUREG/CR–4551, Volume 4) have indicated that the mean core damage frequency from fires is in the range of 10^{-5}/RY and from Seismic events 10^{-6}–10^{-4}/RY.

A common finding from these PRAs is that support system failures have been identified as significant contributors to the probability of core melt. At the support system level, there is often sharing and interconnection between redundant trains, questionable separation and spacial independence between trains, and poor overall general arrangement of equipment from a safety viewpoint. For example, many plants have redundant trains of equipment sitting side by side in a common area and adequate physical separation and protection of redundant safeguard trains is lacking. This type of general arrangement of equipment creates vulnerabilities in that single events such as a fire or a flood can disable multiple trains of safety related equipment resulting in an inability to cool the plant.

Table 2 provides a list of the specific vulnerabilities found from some of these studies and potential modifications to address these vulnerabilities. In general, external event vulnerabilities were identified in:

a. Electrical switchgear/battery failures due to seismic excitation.

b. Water storage tank (CST, RWST) failures due to seismic excitation.

c. Pump and valve common-mode failures (AFW, CCW, SWS, HPIS, LPIS, etc.).

d. Fires in cable spreading rooms, switchgear rooms, or common cable run areas (BWR/PWR).

Modifications include both hardware and procedural components which, if implemented, would serve to reduce the estimated core damage frequency. Table 3 provides an example list of the plant-specific modifications that have been made. These modifications were made based on the insights gained from plant-specific PRAs. However, many fixes are made in the course of doing a PRA which are never quantified or reported. For example, deficient equipment anchorages were found at almost every plant during seismic walkdowns.

Table 1 Summary of PRA Results of Core Damage Frequency (1E–5)

Plant	Total	Total Int'l	Total External	Seismic	Fire	Int'l Flood	Extn'l Floods	High Winds	Light- ning
PWR									
Pt. Beach	31.3	13.9	17.4	6.1	3.3	7.7	–	0.4	0.006
Turkey Pt.	23.6	7.1	16.51	.7	7.5	–	4.6	2.4	0.26
St. Lucie	7.44	1.4	6.04	1.3	4.4	–	0.32	–	0.02
ANO 1	17.9	8.8	9.15	7.3	0.58	–	0.72	0.53	0.02
Mean	20.1	7.8	12.3						
IP2	43.5	6.0	37.5	14.0	19.2	–	–	4.3	–
IP3	15.7	9.0	6.7	0.31	6.3	–	–	0.13	–
Zion	34–40	34.2	0.1–6	<0.1–6					
MS3	15–23	14.7	0.8–8Est	–	–	–	–	–	
Oconee	15–28	7.4	8–21	6.0	1.0	–	0	1.3	–
					10.0 (NRC)	–	2.5 (NRC)	2.3 (NRC)	–
Mean	25–30	14.3	11–16						
BWR									
Quad Cities	19.7	9.9	9.8	8.3	1.3	–	0.01	0.01	0.2
Cooper	43.7	28.9	14.8	8.1	1.1	–	5.	0.4	0.2
Limerick	9.2	8.4	0.8	0.5	0.3	–	–	–	–
Shoreham	7.4	5.4	2.0	–	–	2.0 (NRC)	–	–	
Mean (4) 27	20.2	6.8	–	–	–				

IP2/IP3—Indian Point 2/Indian Point 3
MS3—Millstone 3

usually strengthened, however, and are rarely reported specifically in the PRAs in terms of the impact on CDF or averted risk. Based on the insights gained over the last ten years, almost every systematic examination has resulted in plant-specific insights, that in conjunction with the plant specific evaluation of risk reduction options, would always result in identifying cost effective remedies.

3. Value-Impact Assessment

The analyses performed for the resolution of USI A–45, Decay Heat Removal (DHR) Requirements (Ref. 6), were used to make reasonable estimates of the value of conducting the IPEEE. Specifically, reduction in core damage frequency resulting from proposed modifications and the cost of those modifications were evaluated. It should be noted, however, that the purpose of the USI A–45 program was to evaluate the adequacy of the DHR function only; accident sequences that did not involve this function are not included in the analyses. These excluded sequences involving large LOCAs, reactor vessel ruptures, the pressurized thermal shock sequence, interfacing system LOCAs, and anticipated transients without scram (ATWS). Thus, the core damage frequencies derived under that program do not represent the total frequencies for those operating plants. Including the contri-

butions from those excluded events would result in higher estimated core damage frequencies.

For this study, the following modifications were considered as possible means of reducing the vulnerabilities which would most likely be uncovered in a plant specific IPEEE. They were used here for the purpose of assessing the value-impact of the IPEEE program; however, they do not necessarily represent the only means for reducing plant risk to severe accidents.

(1) Seismic Resistance of Batteries and Switchgear

   a. Ensure that battery installation racks meet current seismic requirements. All racks should be steel with appropriate tiedowns to prevent motion under seismic excitation.

   b. Provide additional ties to floor for electrical equipment (transformers, switchgear, buses, battery chargers, and motor control centers) for anchorages to prevent cabinet motion during seismic acceleration. For tall cabinets, provide additional restraints to prevent toppling.

Table 2 Modification Options Identified for the Case Studies

Plant	Vulnerability	Modification
Pt. Beach	RWST failures and electric switchgear failures from seismic events	Provide water from spent fuel pool and add restraints to switchgear and batteries
	Service water pumps lost from failure due to spray	Install shield wall to protect pump motors
	Loss of safety systems due to fire in CSR and AFW rooms	Install added fire suppression
Turkey Pt.	Surge floods safety systems	Increase height of existing flood wall
	Loss of cooling due to loss of water tanks and CCW heat exchangers from seismic event	Increase strength of tanks and heat exchanger supports
	Loss of safety systems due to fire in CSR	Install additional suppression in CSR
St. Lucie	Loss of safety systems due to CSR fire	Enclose one train of safety-related cables in fire barrier
	Loss of cooling due to loss of water tank	Increase strength of tanks with addition of external supports
ANO 1	Loss of cooling due to failure of EFWS pump and to take water from CST	Install provisions to power auxiliary feed pump from Class IE bus
	Loss of safety systems due to fire in CSR	Add redundant deluge valve with separate sensing and control
	Loss of cooling due to loss of tanks and emergency electric power due to seismic event	Strengthen tanks with external supports and anchor switchgear
Quad Cities	Loss of decay heat removal due to fires in CR or CSR	Enhance operating procedures for the safe-shutdown pump
	Loss of electric power due to seismic events	Upgrade battery racks and add restraints to SWGR and buses
Cooper	Loss of safety systems due to fire in CSR	Add fire barrier around HPCI and RBSW power cables
	Loss of cooling due to failure to tanks and heat exchangers from seismic event	Install added anchorage or tanks and heat exchangers
	Loss of emergency electric power due to seismic events	Add supports and tiedowns to switchgear and transformers
Cooper Alt.	Loss of cooling due to seismic events	Strengthen HTEX mounts, valve, CST, and transformer tiedowns
	Loss of decay heat removal due to floods	Develop procedures for safe shutdown in high flood crests

(2) Seimsic Resistance of Tanks

Upgrade anchorages and walls for water storage tanks (RWST and CST) designed using the procedures of TID 7024 and with H/D ratios greater than 1.

(3) Fire Protection

Where safety-related cabling is concentrated, ensure that adequate fire protection is provided by installation of additional suppression systems, thermal protection, etc. and reliable alternate shutdown capability is available. Review all procedures to ensure that minimal quantities of fuels are present in fire-susceptible areas (control rooms and cable spreading rooms in particular).

3.1 Analysis of Specific Modifications

Table 3 summarizes for selected plants the value-impact analyses resulting from application of the specified system modifications described above. Besides value-impact, core damage probability, population and occupational doses, and costs are shown explicitly in the table. As expected, the value (Col. 4, 5, & 6) and the impact (Col. 7 & 8) of any given modification are plant and site dependent. None of the suggested modifications is cost effective (Col. 9) based on avertible offsite costs alone. However, some modifications may be cost effective if onsite costs are included (Col. 10).

3.2 Plant Specific Value-Impact Analyses

Table 4 summarizes the results of the plant-specific value-impact analyses performed for USI A–45. Various combinations of modifications were evaluated for each plant. Besides value-impact, core damage probability, population and occupational doses, and costs are shown explicitly in the table. As expected, the value (Col. 3, 4, & 5) and the impact (Col. 6 & 7) of any given modification are plant and site dependent. None of the modifications is cost effective (Col. 8) based on avertible offsite costs alone. However, some modifications are cost effective if onsite costs are included (Col. 9).

4. Discussion and Conclusion

As can be seen from Table 1, external events can be significant contributors to overall risk from a nuclear power plant. Previous risk analysis of external events have always uncovered items which were modified to reduce risk (Table 3). In many cases the reduction in risk resulting from these modifications was not quantified and thus value-impacts were not calculated. However, from the data available from the A–45 analysis (Tables 4 and 5) and Table 3 the following conclusions can be drawn:

- the cost of the modifications considered may range from approximately 50K to 24M dollars per plant

- the risk-averted (on-site and off-site) may range from approximately 50 person-rem to 2600 person-rem per plant over the life of the plant

- low-cost fixes may be found to reduce risk and may be cost effective

- for those fixes that were made at specific plants, the averted risk was significant.

- in many cases, even if the cost of doing the IPEEE (estimated to be as much as $1 million at the upper bound) were added to the cost of doing the modifications, the modifications might still be cost effective.

Thus, the staff concludes that there is a high likelihood that conduct of the IPEEE will result in the identification of vulnerabilities that, if fixed, would result in a substantial increase in safety and that could be fixed in cost-effective manner. Accordingly, the systematic examination of each operating nuclear plant could provide the most complete compilation of data and analysis available to develop an integrated perspective on risk from external events. It could also identify human, procedural, design, and operation vulnerabilities and could provide practical means to explore and select cost-effective alternate solutions to plant vulnerabilities. Therefore, a plant-specific examination, like IPEEE, conducted by analysts with access to plant data and procedures, could better establish the level of risk and identify cost-benefit improvements at a particular site.

5. References

1. Generic Letter 88–20, Supplement 4, Individual Plant Examination of External Events (IPEEE) for Severe Accident Vulnerabilities, Draft for Comment, July 23, 1990.

2. "Policy Statement on Severe Accidents," U.S. Nuclear Regulatory Commission, *Federal Register*, Vol. 50, 32138, August 8, 1985.

3. NUREG/CR–3568, A Handbook for Value-Impact Assessment, PNL–4646, dated December 1983.

4. NUREG/CR–5042, Evaluation of External Hazards to Nuclear Power Plants in the United States, LLNL, December 1987.

5. NUREG/CR–5042, Supplement 1, "Evaluation of External Hazards to Nuclear Power Plants in the United States, Seismic Hazards" LLNL, April 1988.

6. NUREG–1289, "Regulatory and Backfit Analysis: Unresolved Safety Issue A–45, Shutdown Decay Heat Removal Requirements, November 1988.

Table 3 Examples of Averted Risk from PRA Experience

Plant	Description of Modifications	Reference
Oconee	Changes to turbine bldg., control room, turbine bldg. eq., and procedures mods to reduce plant vulnerability to internal floods (CCW)	NSAC PRA
Yankee Rowe	Establish risk basis for external event requirement resolution	Chapman
	Tornado/high wind requirement	
	Seismic design changes	
Indian 2	Mod. of structural design of control room	IP2 PRA
Millstone 3	Replace diesel generator oil cooler anchor bolts (seismic)	MS3 PRA
Conn Yankee	App. R Mod. Tornado/high wind mod.	
Pt. Beach	Add additional fire suppression	

Table 4 Application of Specified System Modifications; Results of Value-Impact Analyses for Specific Plants

Plant (1)	Var. No. (2)	Base p(CDF) (per r–yr) (3)	dp(CDF) w Var. (per r–yr) (4)	Averted Dose (person-rem) Offsite (5)	Net Onsite (6)	Averted Impact (Gross) ($xE6) (7)	Onsite Costs ($xE6) (8)	Value-Impact ($/person-rem) GrossNet (9)	(10)
Pt. Beach	3	3.13E–4	1.2E–5	36	15	0.99	0.26	2.8E4	1.4E4
	1b		1.5E–5	45	18	0.24	0.33	5.3E3	=0
Turkey Pt.	3	2.36E–4	7.2E–5	535	81	3.10	2.33	5.8E3	1.3E3
	2		1.3E–5	99	15	0.91	0.42	9.2E3	4.3E3
St. Lucie	3	7.44E–5	2.9E–5	100	37	0.60	1.05	1.05E4	=0
	2		1.2E–5	42	15	0.052	0.44	1.2E3	=0
ANO 1	1b/2	1.79E–4	6.4E–5	84	71	0.131	1.97	1.6E3	=0

Notes:
Column 2,   Modifications are as described in Section 3.
Column 6 = Averted Onsite Dose—Installation Dose.
Column 7 = Present Worth of Installation Costs + Operation and Maintenance Costs + Replacement Power Costs During Installation + Cost of Limited-Scope PRA.
Column 8 = Present Worth of Replacement Power Costs + Loss of Investment + Cleanup Costs.
Column 9 = Col 7/Col 5.
Column 10 = (Col 7 – Col 8)/(Col 6 + Col 5)

Table 5 Modifications Based on Limited–Scope PRA, Results of Value-Impact Analyses for specific plants

Plant (1)	p(CDF) Base (per r–yr) (2)	dp(CDF) w Var. (per r–yr) (3)	Averted Dose (person-rem)		Averted Impact (Gross) ($xE6) (6)	Onsite Costs ($xE6) (7)	Value-Impact ($/person-rem)	
			Offsite (4)	Net Onsite (5)			Gross (8)	Net (9)
Pt. Beach	3.13E–4	2.7E–5	81	33	1.23	0.59	1.5E4	5.6E3
Turkey Pt.	2.36E–4	8.5E–5	634	96	4.0	2.75	6.3E3	1.7E3
St. Lucie	7.4E–5	4.1E–5	144	51	0.65	1.49	4.5E3	=0
ANO 1	1.79E–4	6.4E–5	84	71	0.131	1.97	1.55E3	=0
Quad Cities	1.97E–4	9.11E–5	2521	103	5.94	2.72	2.4E3	1.2E3
Cooper	4.37E–4	3.01E–4	2295	278	24.3	6.58	1.1E4	6.9E3
C (alt.)		2.95E–4	2241	271	3.19	6.42	1.4E3	=0

Notes:
Column 5 = Averted Onsite Dose – Installation Dose
Column 6 = Installation Costs + Operation and Maintenance Costs + Replacement Power Costs During Installation in 1985 Dollars
Column 7 = Present Worth of Replacement Power Costs + Loss of Investment + Cleanup Costs
Column 8 = Col 6/Col 4
Column 9 = (Col 6 – Col 7)/(Col 5 + Col 4)

APPENDIX 4
DOCUMENTATION

This appendix provides the guidelines for documentation and reporting format and content for the IPEEE submittal. The major parts of this appendix are the guidelines for seismic analysis (Section 4.2), internal fire analysis (Section 4.3), other analyses (Section 4.4). Licensees are requested to submit their IPEEE reports using the standard table of contents given in Table C.1 of NUREG-1407 or provide a cross reference. This will facilitate review by the NRC and promote consistency among various submittal. The contents of the elements of this table are discussed further below.

The level of detail needed in the documentation should be sufficient to enable the NRC to understand and determine the validity of key input data and calculation models used, to assess the sensitivity of the results to all key aspects of the analysis, and to audit any calculation. All important assumptions should be reported. It is not necessary to submit all the documentation needed for such an NRC review. Relevant documentation should be cited in the IPEEE submittal, and be available in easily retrievable form. The guideline for judging the adequacy of retained documentation is that independent expert analysts should be able to reproduce any portion of the results of the calculations in a straight forward, unambiguous manner. To the extent possible, the retained documentation should be organized along the lines identified in the areas of review. Any information that is comparable to that provided under the IPE for internal events can be incorporated by reference.

4.1 General

4.1.1   Conformance with Generic Letter and Supporting Material

Certification should be provided that an IPEEE has been completed and documented as requested. The certification should also identify the measures taken to ensure the technical adequacy of the IPEEE and the validation of results.

4.1.2   General Methodology

An overview description of the methodology employed in the IPEEE for each external event examined should be provided.

4.1.3   Information Assembly

Reporting guidelines include:

1. Plant layout and containment building information not contained in the Final Safety Analysis Report (FSAR).

TABLE 3.2

REVIEW LEVEL EARTHQUAKE - WESTERN UNITED STATES PLANT SITES

0.5g*

Trojan	Rancho Seco
Washington Nuclear	Palo Verde

Seismic Margin Methods Do Not Apply To the Following Sites:

Diablo Canyon	San Onofre

NOTES:

* Indicates a Western United States site whose default bin is 0.5g unless the licensee can demonstrate that the site hazard is similar to those sites east of the Rocky Mountains that are found in the 0.3g bin.

  Changes in the review level earthquake from 0.5g to 0.3g should be approved prior to doing significant analysis.

NRC FORM 335
(2-89)
NRCM 1102,
3201, 3202

U.S. NUCLEAR REGULATORY COMMISSION

BIBLIOGRAPHIC DATA SHEET

(See Instructions on the reverse)

1. REPORT NUMBER
(Assigned by NRC, Add Vol., Supp., Rev., and Addendum Numbers, if any.)

NUREG-1407

2. TITLE AND SUBTITLE

Procedural and Submittal Guidance for Individual Plant Examination of External Events (IPEEE) for Severe Accident Vulnerabilities

Final Report

3. DATE REPORT PUBLISHED

MONTH	YEAR
June	1991

4. FIN OR GRANT NUMBER

5. AUTHOR(S)

J. T. Chen, N. C. Chokshi, R. M. Kenneally, G. B. Kelly, W. D. Beckner, C. McCracken, A. J. Murphy, L. Reiter, D. Jeng

6. TYPE OF REPORT

7. PERIOD COVERED (Inclusive Dates)

8. PERFORMING ORGANIZATION – NAME AND ADDRESS (If NRC, provide Division, Office or Region, U.S. Nuclear Regulatory Commission, and mailing address; if contractor, provide name and mailing address.)

Division of Safety Issue Resolution
Office of Nuclear Regulatory Research
U.S. Nuclear Regulatory Commission
Washington, DC 20555

9. SPONSORING ORGANIZATION – NAME AND ADDRESS (If NRC, type "Same as above"; if contractor, provide NRC Division, Office or Region, U.S. Nuclear Regulatory Commission, and mailing address.)

Same as above

10. SUPPLEMENTARY NOTES

11. ABSTRACT (200 words or less)

Based on a Policy Statement on Severe Accidents, the licensee of each nuclear power plant is requested to perform an individual plant examination. The plant examination systematically looks for vulnerabilities to severe accidents and cost-effective safety improvements that reduce or eliminate the important vulnerabilities. This document presents guidance for performing and reporting the results of the individual plant examination of external events. The guidance for reporting the results of the individual plant examination of internal events (IPE) is presented in NUREG-1335.

12. KEY WORDS/DESCRIPTORS (List words or phrases that will assist researchers in locating the report.)

Severe Accidents Policy Statement
Individual Plant Examination of External Events (IPEEE)
Vulnerabilities
IPEEE Guidance

13. AVAILABILITY STATEMENT
Unlimited

14. SECURITY CLASSIFICATION
(This Page)
Unclassified
(This Report)
Unclassified

15. NUMBER OF PAGES

16. PRICE